Managing Indoor Air Quality
Fifth Edition

MANAGING INDOOR AIR QUALITY
FIFTH EDITION

H.E. BURROUGHS, CIAQP
SHIRLEY J. HANSEN, PH.D.

THE FAIRMONT PRESS, INC.

CRC Press
Taylor & Francis Group

Library of Congress Cataloging-in-Publication Data

Burroughs, H.E.
    Managing indoor air quality / H.E. Burroughs, Shirley J. Hansen. -- 5th ed.
      p. cm.
    Authors' names in reverse order in some earlier eds.
    Includes bibliographical references and index.
    ISBN-10: 0-88173-661-9 (alk. paper)
    ISBN-10: 0-88173-662-7 (electronic)
    ISBN-13: 978-1-4398-7014-3 (Taylor & Francis : alk. paper)
    1. Indoor air pollution. 2. Air quality management. 3. Buildings--
      Environmental engineering. I. Hansen, Shirley J., 1928- II. Title.

    TD883.1.H36 2011
    697--dc22

                                      2011003466

Managing indoor air quality, fifth edition/H.E. Burroughs, Shirley J. Hansen.

Published by The Fairmont Press, Inc.
700 Indian Trail
Lilburn, GA 30047
tel: 770-925-9388; fax: 770-381-9865
http://www.fairmontpress.com

Distributed by Taylor & Francis Ltd.
6000 Broken Sound Parkway NW, Suite 300
Boca Raton, FL 33487, USA
E-mail: orders@crcpress.com

Distributed by Taylor & Francis Ltd.
23-25 Blades Court
Deodar Road
London SW15 2NU, UK
E-mail: uk.tandf@thomsonpublishingservices.co.uk

Printed in the United States of America
10 9 8 7 6 5 4 3 2 1

ISBN-10: 0-88173-661-9 (The Fairmont Press, Inc.)
ISBN-13: 978-1-4398-7014-3 (Taylor & Francis Ltd.)

CONTENTS

Management by Complaint • Developing an IAQ Program • Policy • The IAQ Manager • Management Plan • IAQ Resource Notebook • Management Concerns • Economic Factors • Legal Matters—Specific Policy Issues • Securing Consultant Services • Education & Training • Communication in IAQ Management

From the Primary Revision Author

This is the fifth edition of *Managing Indoor Air Quality* spanning the last 20 years. This alone is obvious proof that indoor air quality represents an important, growing, and changing issue. And yet, even in the face of this maturing technology, we face another set of quick and glib labels that are challenges to the building owner/manager...

"Sustainability" and the "Net-zero Energy Building." It is my fear that in the passionate pursuit of these positive and lofty goals that, in their name, we will repeat the sins of the past three decades. That is when we created unacceptable indoor environments (Tight Building Syndrome) and simultaneously funded retirement programs for the legal profession—all in the name of "Energy Management."

Upon further reflection, however, it is my sense that unlike in the 70s and 80s, IAQ is now more deeply engrained and integrated into the design/construct/manage building dynamic. I confess that we can unfortunately credit the legal profession for some of that progress. As testimony of this acceptance and integration, I give due credit to the work of ASHRAE in the development of their recently published guidance document, the Indoor Air Quality Design Guide. This document was sponsored by EPA and developed by a multi-disciplinary team of volunteers (of which I am proud to have been a part) that represented the architecture, engineering, construction, IAQ practitioners, commissioning, and building operation perspectives. The body of their work is primarily in digital form and is a comprehensive, thorough, and detailed compendium of advanced best practices regarding the attainment and sustainment of acceptable indoor air quality.

Admittedly, there are voids in our knowledge base about IAQ, especially in the areas of human response; related adverse health effects in the present of trace airborne contaminants; and the interrelation of IAQ influencing factors. However, as evidenced by the ASHRAE IAQ Design Guide, we already know a great deal about creating and managing buildings that demonstrate acceptable indoor environments; that are sustainable with minimal ecological impact; and employ effective energy management tactics. The goal is to simply apply better and more consistently what we already know how to do.

The message to the building owner is to be familiar with fundamentals of IAQ and to be willing to consider and apply the benefits of, for example: integrated design; life-cycle cost decisions; enhanced filtration; commissioning; advanced ventilation strategies; moisture management; and thorough operation and maintenance practices.

Under the masterful initiation of Shirley Hansen, author of the first edition of *Managing Indoor Air Quality*, it has been the constant goal of this book to provide the building owner/manager with those fundamentals of IAQ for both the avoidance as well as coping with IAQ issues. My further thanks to Ian Cull for his input on updates and edits.

H.E. Barney Burroughs, Principle Revision Author

PREFACE

As predicted in the earlier editions, concerns over Indoor Air Quality (IAQ) have driven it to the forefront of current workplace problems. More than ever, health, economic and legal matters associated with IAQ seem destined to make it a dominant problem for developers, owners and managers of commercial and institutional properties well into this millennium. Indoor air problems have definitely made the engineer's and facility manager's jobs more difficult and put the owner at greater risk.

We have become more conscious of human concerns and have become more aware of the cost associated with achieving indoor air quality. Employers are being plagued by "sick buildings," associated absenteeism, and lower productivity. Elements of risk and liability associated with poor IAQ have caught the attention of our litigious society. Thus, employee and occupant lawsuits continue.

The IAQ problem is real and it is not going to go away. The ASHRAE standard 62, Ventilation for Acceptable Indoor Air Quality, and its future revision process will continue to force us all to consider the indoor environment even in "healthy" buildings. This book is designed to help those who are responsible for *managing* that environment. It encompasses the whole range of contaminants, thermal conditions and other factors that contribute to, or detract from, occupant comfort and a productive workplace. Focusing on IAQ issues from a management perspective, this book is not intended to be a detailed, technical book. However, enough detail is provided for it to be a practitioner's handbook designed to help owners, managers, operators or anyone responsible for operating and maintaining a facility. While not directed primarily at the design professionals, those who offer owners technical and financial support will find it beneficial to examine the IAQ issue from management's point of view.

Managing Indoor Air Quality is structured as a guide as well as a reference document to treat existing indoor air problems effectively, and to help prevent costly IAQ problems from occurring. Finding solutions to IAQ problems is often a complex, multifaceted, multidisciplined endeavor. A single discipline approach from the environmental engineer, the industrial hygienist or the medical doctor, unfortunately tends to narrow the control and treatment options. This book cuts across these professions without the specificity and bias of any one discipline, to offer those concerned with the total facility a broader approach.

For all of the progress we have made in recent years, there are still large gaps in IAQ knowledge. The issue, however, is no longer new and experiences in almost every type of facility exist to guide us. This revised edition is rich in IAQ history and new practical, pragmatic suggestions throughout its procedural guidelines.

The reader is cautioned that every IAQ situation is unique and the procedures discussed herein offer only general guidelines. When problems persist, it is always prudent to call in an IAQ professional, special IAQ consultant and/or team to examine a specific situation. Reference herein to any specific commercial product, process, or service by trade name, trademark, manufacturer, or otherwise, does not necessarily constitute or imply its endorsement, recommendation or favoring by the authors.

As IAQ concerns have erupted around us, the means to identify, control and treat a myriad of pollutants has mushroomed. Researching, sifting and sorting this burgeoning data was a mammoth task made easier by many people. I would like to express my continuing appreciation to those who made the first edition possible, and my thanks to the many readers who have taken the time to tell me of the value the original text was to them. It is very gratifying to know it served so many so well; however, in the rapidly changing world of IAQ a book of this nature must be revised and updated to remain useful.

Since my professional attention has shifted away from IAQ in recent years, I was absolutely delighted when H.E. (Barney) Burroughs agreed to take on the revision process. A first book, like a first form, holds a special place. You can't turn a "first born" over to just anyone. As you read through the contributions Barney has made, you'll see the care, integrity and professionalism that all of us in the industry associate with him. He's made a good book much, much better. But then I knew he would. Thanks, Barney.

Shirley J. Hansen

Chapter 1

INDOOR AIR QUALITY: AN OVERVIEW—WHERE ARE WE?

Through the years, man has built increasingly elaborate boxes to protect himself from the elements. Designed to keep out the rain and snow, warm him in the winter and cool him in the summer, he now lives and works inside these boxes up to 90 percent of the time. Rather than hold environmental hazards at bay, however, he has trapped himself inside in a chemical stew of contaminants that might make him sick, even kill him.

Buildings don't always protect their occupants from pollution. Just the opposite, the molds, fungi, dust and toxic gases trapped or growing on the inside may well exceed those outdoors. By shielding ourselves from the outside environment, we have created an inside environment with a whole new set of problems. To understand this delicate and fragile interrelationship and interdependence between the indoor and the outdoor environment, building owners and managers need to view the building as a "habitat" or an "ecosystem."

THE BUILDING AS A HABITAT

To view the ecological factors of the building as a habitat is to understand the influences and attributes of both the indoor and the outdoor environment. This includes the air within the conditioned and controlled space where we breathe and work. It includes the complete air path and the systems to which it has been exposed—is it clean and dry, void of contamination, and free of microbial growth and pathogens. It includes the building envelope—is it sound, visually pleasing, and well sealed and dry, allowing in light without water. It includes the building site and its setting—the role of greenery and landscaping, natural lighting and noise levels, as well as odor or pollutant sources. Lastly, it includes the impact of the building habitat upon the global environment—its energy demands, pollutants, and waste products (air, solids, and waste water), demands upon raw material resources, and even the consumption of financial capital.

To this delicate ecological relationship, we, the occupants, are added. We are both the violators and victims of our habitat. Our very presence affects levels of temperature, relative humidity, odor, metabolic gases, volatile organic and inorganic gases, and particulate matter. We shed potential disease-bearing pathogens and create a constant stream of waste effluent from our metabolic and biological functions. Our activities and related equipment and furnishings further confound our environment with chemicals and particulates.

This complex IAQ ecosystem model, therefore, consists of the occupants, their activities, the air pathway and the HVAC system, the building envelope, and its environmental setting. One additional component undergirds this matrix—time. Time is essential to our understanding of the causes and prevention of poor indoor air quality. Systems, situations, loads, usage, efficiencies, and costs all change over time. As buildings age, they wear, deteriorate, leak, break, and fail. This process of continuous degradation makes maintenance and operating practices a major component of this IAQ model.

It is the goal of this book to bring about an understanding of these complex and delicate relationships and their interdependence. Thus, we cannot deal with indoor air quality causes singularly without understanding all aspects of the interlinking model. Much like the proverbial house of cards, it takes all the cards to attain and maintain the structural integrity, but only a single card can bring down the structure. This concept is nowhere better expressed than in the holistic and integrated design approach that is promulgated in the recently published ASHRAE advanced IAQ design guide *IAQ Guide: Best Practices for Design, Construction, and Commissioning* (2009).

WHEN BUILDINGS BECOME SICKENING

It usually starts with building occupant's complaints of discomfort, headaches, nausea, dizziness, sore throats, dry or itchy skin, sinus congestion, nose irritation or excessive fatigue. Suddenly the words that are rapidly becoming the bane of building owners, facility managers and engineers are declared: Sick Building Syndrome (SBS)! Of course the building is not really sick, but it has developed a condition that can make its occupants uncomfortable, irritated, or even ill. A building is generally deemed as *sick* when the building's occupants exhibit such symptoms and the complaints persist for more than two weeks—particularly, if the symptoms disappear when the sufferers leave the building for the weekend.

The term "syndrome" is used by the medical profession to indicate a cluster of symptoms occurring together. The exact relationship existing among those symptoms may be unknown. In a sick building, symptoms experienced by occupants are those often associated with the upper respiratory tract or they may manifest themselves as headaches, dizziness, nausea, lethargy and fatigue. Other symptoms are revealed in areas where the body is directly exposed, such as dry or itchy skin and eye irritation. A sick building can also aggravate existing illnesses.

An associated problem is Building Related Illness (BRI). BRI is deemed as a building associated, clinically verifiable and diagnosable disease. If signs of actual illness are present and can be attributed to a condition in the facility, the building can be classified as BRI.

The key difference between SBS and BRI is that specific SBS contaminants may not be known. SBS is diagnosed when complaints and symptoms are clearly associated with building occupancy, but no causal agent can be positively identified. Complaints are often resolved by increased ventilation, by more effective controls/substitutions of possible sources and by improved maintenance. Often, BRI is an advanced stage of SBS. The dirt, dust, moisture and stagnant water typical of the poor maintenance that causes SBS is an ideal home for the BRI-causing bacteria looking for an amplification site. Thus, it is unusual that a building will get to BRI stage without first going through SBS with its related comfort complaints.

WHY NOW? WHAT IS DRIVING THE IAQ ISSUE?

What has made IAQ the ongoing hot topic into the first decade of the century? It is important to note that as the topic has entered the 21st century, IAQ has become the broader topic of IEQ, which also recognizes the role of sound and light as contributing factors.

Since the higher energy costs of the 1970s, new construction and building modifications brought tighter envelopes and increasing reliance on mechanical ventilation. By reducing the intake of outside air, pollutants that were already there have been concentrated and their effects on humans have become more obvious. Energy cost has remained a significant driver of IAQ related issues. At the same time, construction materials and furnishings, such as carpeting and work stations, have brought more contaminants into homes and the work place. Technological changes have made copiers, printers, computers, facsimile machines, etc. common place in the office. Operating and cleaning this equipment have brought more pollutants into the office. They have also caused dramatic changes in office procedures prompting ergonomic and organizational stress problems, which have seemed to heighten health problems associated with indoor air pollution.

Against this backdrop, greater press coverage has raised public awareness of indoor contaminants. Furthermore, the federal government got involved early on. Under the Superfund Bill (P.L. 99-499), Congress authorized the Environmental Protection Agency (EPA) to research radon and this invoked EPA interest and involvement in the indoor environmental arena. And in addition, OSHA announced in April of 1994 the intention to regulate indoor air quality and invited public comment. They were inundated with oral and written comments which delayed formal regulation for the balance of the decade and brought about a final withdrawal of the intent in 2002. Although the document was labeled as an IAQ regulation, the primary focus was the limitation of ETS (environmental tobacco smoke) in the workplace. Even though the regulation under was never promulgated, most buildings responded with self-imposed voluntary smoking limitations. Many states and local jurisdictions followed and continue to promulgate local regulations on smoking in public space(s).

IAQ is costly business. U.S. Army research findings have suggested that IAQ related health problems are costing an estimated $15 billion a year in direct medical costs and about 150 million lost workdays. Garibaldi and Dixon's review of the Army data lead them to conclude that at least $59 billion of indirect costs could be added to the price tag. In 1988, *Time* reported an out-of-court settlement of close to $600,000 had been made to an occupant in a new office building in Goleta, California, who claimed to have lost consciousness and suffered permanent brain injury due to formaldehyde fumes. In more recent judgments involving several public courthouses, costs of litigation and mitigation have exceeded the original cost of the structures.

Concerns regarding health, productivity, absenteeism, vacant facilities and the threat of lawsuits are destined to make indoor air quality (IAQ) the dominant issue for building owners and facility operators into the new millennium. According to one study by Dorgan, the cost of poor IAQ in the hospitality industry alone exceeds $19 billion. This estimate includes absenteeism and sick leave, health-related costs, turnover, and productivity improvement.

Early in this decade, Fisk published an estimate of potential savings and productivity gains to US industry by providing a better indoor environment. He estimated gains of from $18 to $43 billion due to reduced respiratory disease, allergies, and asthma. In addition, he estimated another $12 to $125 billion from direct improvements in worker performance unrelated to health. He reported that sample calculations indicated that the potential financial benefits of improving the indoor environment exceed costs by a factor of 18 to 47 as the return on investment. As a part of the report, the costs of building-influenced adverse health-care were estimated at $35 billion.

In a more recent 2005 study, Levin evaluated the U.S. national cost of IAQ expenditures under contract with EPA. He reviewed a number of categories including: consulting services; building diagnostics and related laboratory services; building remediation and renovation driven by IAQ; additional design and construction cost; air duct cleaning services; improved filtration cost; litigation and insurance; and additional specialty staff costs. The study reported total IAQ related expenditures of approximately $15.9 billion. Further, this estimate did not include related expenditures related to: significant reduction of contaminant source strength driven by IAQ concerns; additional construction and operating costs for attaining higher standards of care; cost related to the reduction of smoking in public space; staff costs relating to IAQ complaint response; and training certification costs.

HUMANE CONCERNS
BECOME PUBLIC HEALTH CONCERNS

Since early in this century, episodes of extreme air pollution in highly industrialized and urbanized locations have focused attention on the quality of the air we breathe and associated environmental health concerns. Studies of such episodes and demonstrated increased incidence of respiratory disease and mortality in populations exposed to such pollutants as nitrogen oxide, sulfur oxide and particulates, such as coal dust, and prompted laws to establish specific ambient air quality standards designed to protect the public's health from this type of pollution.

One study of outside air and associated health effects by the Harvard School of Public Health discovered indoor air added another dimension in determining exposure levels. In its Six Cities Study, Harvard researchers found children 6 to 10 years of age living with cigarette-smoking parents had more respiratory illnesses and impairment of lung functions than children living with non-smokers. In work published by EPA in 2001, the percentage of children with asthma has doubled in two decades, rising from 3.6 percent in 1980 to 8.7 percent (6.3 million children by 2001). Researchers suspect exposure to dust miles, cockroaches, pesticides, tobacco smoke, ozone and soot. In a similar study by the University of Southern California Medical School, it was determined that asthma increased from 50 percent to 4 to half times in children one to five years old when exposed to similar indoor contaminants of concern.

The Harvard study and others have shown that indoor environments can play a critical role in exposure levels and health outcomes. Thus, it has become increasingly evident that the levels of many pollutants may be higher inside buildings than outside.

More Than Just Numbers

Reducing the misery, illness and loss of life to mere statistics risks diminishing the human suffering experienced by individuals due to indoor air pollutants. A few numbers, however, can help document the broad impact air quality has on our lives. In 1987, Platts-Mills estimated 15 percent of the admissions to hospital emergency rooms each year, or 500,000 patients, could be attributed to dust mites and other airborne allergens. According to EPA statistics, asthma is the leading cause of childhood hospitalizations and school absenteeism, accounting for 100,000 child hospital visits a year, at a cost of almost $2 billion, and causing 10 million school days missed each year. The most sensational health problems are associated with IAQ, Legionnaires' disease, continues to plague facility managers. Writing in the late 1980s, Fang reported that the *Legionella* species of bioaerosols, (from cooling towers, hot-water systems and even hot-tubs in residences) account for 8 to 10 percent of community-acquired pneumonia, or 50,000 to 60,000 illnesses, each year. Millar reports up to 8000 deaths annually from Legionella related causes, with many more likely due to misdiagnosis and confusion with other pneumonia varieties. Humidifier fever and hypersensitivity pneumonitis is estimated to occur annually in 1 to 4 percent of the 27 million office workers, running the estimates as high as 1.1 million per year. When health effects attributed to volatile organic compounds (such as formaldehyde), radon, environmental tobacco smoke, etc. are added in, the numbers keep growing. Aside from the human suffering these numbers represent, they suggest economic woes of serious dimensions.

ECONOMIC ISSUES

Corporations with extensive office complexes, have found that approximately 35 percent of their total operational budget for an office facility is devoted to personnel. Actions, which save few cents in other segments of the facility budget but have a negative effect on personnel are, therefore, apt to have a greater negative net effect on the bottom line.

Saving Energy and Losing Money

If reduced ventilation saves 10 percent of a 4 percent energy budget, total costs have been cut .4 percent. If that action should cause a 3 percent increase in absenteeism, there would be a net loss in the total operational budget of .65 percent. That's poor economy. There are, of course, ways to optimize energy efficiency and indoor air quality that will be discussed later. Furthermore, an energy efficient building need not be automatically responsible for IAQ associated problems, if it is managed and maintained properly.

Lost Productivity

We have all observed in ourselves and others that people who are ill do not function as well. They are less productive. While reports of IAQ problems often mention lower productivity, the effects on occupant productivity have not been fully measured and documented; however, it is reasonable to assume that these effects can be very substantial. Isolated studies and logic would suggest that IAQ problems, which have an adverse effect on health, also have a negative impact on productivity.

When office personnel become ill in a given suite of offices and productivity drops, the company moves out. In the greater Washington, DC, area, a 22-story sick building stood empty for months while team after team tried to identify the indoor air pollutants and sources causing the health problems. In Fort Worth, a modern landmark high-rise building stands vacant and boarded up as a victim of massive moisture and mold infestation. These examples represent 0 productivity since there is no cash flow from the investments. Even worse, in the case of the Fort Worth structure, the staggering renovation or demolition costs create negative asset value. Costs associated with a sick building that stands empty can be staggering. Furthermore, it poses an unending nightmare for the owner, architect, engineer and facility manager. Owners cannot afford empty buildings. Companies cannot afford higher absenteeism and lower productivity. However, legal costs may prove to be the greatest economic burden of all.

LEGAL IMPLICATIONS

In today's litigious society, the specter of legal action always looms large. As a harbinger of things to come, the *Wall Street Journal* as early as 1988, reported:

"SICK BUILDINGS LEAVE BUILDERS AND
OTHERS FACING A WAVE OF LAWSUITS

More office workers are filing lawsuits, claiming they were made ill by indoor air pollution from such things as insect sprays, cigarette smoke, industrial cleaners, and fumes from new carpeting, furniture, draperies and copiers."

Most journals for design professionals, (architects, engineers and interior designers) have run articles by attorneys urging them to limit their IAQ liabilities.

As early as June, 1986, *The Professional Liability Perspective*, published by a prominent Northern California insurance broker, warned design professionals, "The potential liability problems posed by indoor pollution are compounded by the fact that the pollution exclusion in your policy of professional liability insurance is all-compassing. It extends to every form of environmental contamination imaginable. The risk is simply not insurable." These early forewarnings have continued to prove out through the '90s as high profile cases have dominated both the trade and popular press. Although details regarding the settlement amounts were sealed by the court, the trial of *Prudential vs. Cole* established significant points of IAQ case law. Specifically, this case resolved that the building owner was responsible and liable for negative health effects experienced by a tenant of the building. Further, it established that the building was a "product" and strict product liability laws would apply. Later, courthouses in Du Page County, Illinois, and Polk and Martin Counties in Florida resulted in "feeding-frenzy" level media coverage and notable monetary judgments. More recently, the litigation has focused on mold related incidents in both commercial and residential properties. A landmark case in Texas concerned the Ballard family who was awarded $32 million against their insurance carrier (later reduced during appeal). The complaint was due to mold related issues in their residence, but the award was rendered because of delay and misconduct by the insurer. The elegant Hilton Kahlia Tower in Honolulu reopened at an estimated cost of diagnostics and remediation in excess of $ 72 million. Litigation is underway in southern Florida against a well-know apartment developer in excess of $78 million in mold related issues in high rise luxury apartments. Ed McMann of TV fame was awarded over $7 million for his aging poodle allegedly affected by improper mold remediation in his dwelling. Although the Ballard ruling was later reduced to several million dollars, these actions are typical of the mold related litigation frenzy. The result is that the insurance industry has excluded mold coverage from virtually all residential and commercial insurance coverage. Related to this outcome are similar exclusions or severe limitations on professional liability and E and O insurance for professionals in the field. Thus, litigation or the threat of litigation is a significant concern and a potent risk for building owners and managers as well as the construction team.

The devastation during and following the Katrina Hurricane experience added fuel to the litigation frenzy. This was due partly to the extensive scope of the primary storm damage and secondary mold damage to buildings over the entire Gulf Coast. But the issue was also fueled by slow and/or advisorial response by some of the insurance carriers. The result was the death knell of coverage by insurers and water/fungal damage in indoor space joined the pollution exclusion in property insurance coverage—giving new meaning to "self-insured." Not only were building owners and managers affected, but the IAQ professionals faced new levels of professional liability costs.

INDOOR AIR POLLUTANTS

What's at the root of the problem? What causes buildings to become "sick?" Indoor air constituents! ...which can affect the health of occupants. These are divided into particles (solids or liquid droplets) and gases or vapors. While tobacco smoke immediately comes to mind as an indoor air pollutant, since it is now heavily regulated in the United States, other constituents that have also been treated extensively in the press, such as asbestos fibers, radon and formaldehyde are readily recognized as contributors to the problem. Allergens, such as pollen and mold, are well known to asthmatics. These components are often common constituents of air. When they attain sufficient concentration to degrade the integrity or quality of the air or to potentially cause health effects, they are referred to as "contaminants" or "pollutants." The relationship of constituent to pollutant is best summarized by the 16th century Swiss physician Paracelius who stated "Everything is a poison. It all depends upon the dosage."

Carbon dioxide (CO_2) is a common constituent of air which is sometimes listed as a pollutant, but does not present a problem unless the concentration level is very high. CO_2 does, however, serve as an excellent surrogate in testing for other gases produced or accumulated during occupancy.

The list of problem or probable contaminating pollutants is long. Cigarette smoke alone is known to contain 4,700 chemical compounds, including several that have been shown to be highly toxic in animal tests, and 43 suspected carcinogenic compounds. This prompted scrutiny by EPA who developed a white paper with particular focus on second hand smoke or "environmental tobacco smoke" (ETS). In addition to establishing ETS as a potential carcinogen and an indoor contaminant of concern for young children, the document provoked the proposed OSHA ruling of '94 and the subsequent tobacco litigation and local jurisdictional regulations. The ruling was buried in negative comment and was never promulgated. However, building owners united with local authorities and reacted with self regulation and generally banned smoking in public spaces. As a result, ETS is no longer as widespread an indoor contaminant of concern. The chapter on classifying IAQ problems discusses all the major indoor pollutants.

Some published papers on IAQ differentiate between contaminants and pollutants; citing a *contaminant* as anything foreign in the air or water and a *pollutant* as something with an adverse effect on humans. At this point, it seems to be a forced distinction. Webster uses pollutant as a synonym for contaminant and the terms are used interchangeably in this book. The important thing for the reader of this book to understand is that IAQ concerns and adverse health effects can be caused by common constituents of air only when they attain abnormally elevated concentrations or long exposure times.

TREATING THE PROBLEM

Nearly every article on indoor air quality will rhetorically mention the energy efficient "tight" building. Since returning to those good ol' leaky buildings does not seem likely, the knee jerk reaction was to recommend an increase in ventilation. Ventilation is a key ingredient, but those who offer ventilation as the only means of treating these contaminants fail to consider the problem in sufficient depth. An August 1988 quote from a trade journal article remains typical of the cursory remedies put forth to resolve a complex problem:

"Practically speaking, attributing the problem to an unidentified contaminant or contaminants rather than saying, simply that the ventilation is inadequate doesn't make much difference if the prescription is to deal with ventilation."

But knowing what the contaminants are and their sources does make a difference! Figure 1-1 shows ACH plotted against radon concentrations for 250 buildings taken from three studies to illustrate the role of ventilation when the source is inside the building. The secondary and equally important effect of increased ACH of ventilation air is that it will also create more positive pressure to act as a barrier to ground gas penetrating the footprint foundation in the first place. However, if the outdoor air introduced as ventilation air is pre-contaminated with constituents of concern, such as carbon monoxide from traffic exhaust, then increased ACH will exacerbate the internal problem.

IAQ CONTROLS

Understanding the ways pollutants can be controlled puts ventilation as a solution in perspective and enlarges upon the maintenance opportunities.

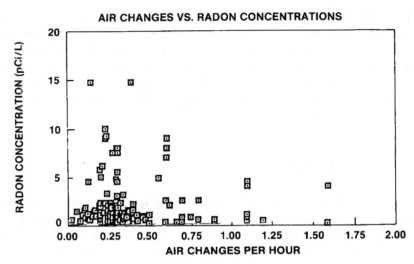

Source: New York State Energy Office based on the work of LBL, W.S. Fleming, Wagner

Figure 1-1. Air Changes vs. Radon Concentrations

There are only three techniques for the control all of indoor air pollutants: dilution; extraction; and source control. The following lists derivatives of these techniques:

1. Elimination, removal or substitution at the source (Source Control);
2. Filtration and air cleaning of airborne contaminants (Extraction);
3. Ventilation of the indoor air with outside air or filtered and conditioned return air (Dilution);
4. Encapsulation, or otherwise interfering with the materials' ability to give off pollutants (Source Control);
5. Isolation by space, distance, or time of use, of a potential contaminant (Source Control; and
6. Education and training of building occupants, especially operations and maintenance personnel.

The EPA, along with other cognizant authorities, consistently stated that source control is the most direct and dependable control option, and the only effective one when strong pollutant sources are present. Source control usually demands some type of investigative procedure to determine precise source components such as the review of MSDS (Material Safety Data Sheet) data.

INVESTIGATION HEADACHES

The greatest "headaches" associated with indoor air problems may come from investigating suspected indoor air quality concerns. Facility managers, directors of maintenance or operations, and plant engineers, trying to diagnose a problem are apt to get the very headaches (for different reasons) they are working to help occupants avoid. Assessing and controlling indoor air pollutants is a complex process. IAQ investigations tend to be complicated by the passage of time, emotional overtones and the fact that symptoms are not easily attributed to a single cause. Furthermore, investigations frequently require a multi-disciplinary approach calling, for example, for both medical and engineering expertise.

The National Institute of Occupational Safety and Health (NIOSH), which has broad experience in IAQ investigations, has observed that "the application of standard industrial hygiene, medical and epidemiological techniques may prove to be inconclusive." Drawing from its experience in over 500 investigations, NIOSH has developed a solution-oriented approach that progressively eliminates the most frequent causes. The original NIOSH approach and other investigative procedures are discussed in the chapter on Investigating Indoor Air Problems.

NIOSH's logical approach can guide the layman's first steps as well. Be practical: if it smells like there's a skunk under the house, chances are there's a skunk under the house. Establishing procedures to remove the most likely causes first just might avoid much of the headache and cost associated with diagnosing the problem. The concept carries over to preventive procedures, especially in the area of maintenance, for the easiest remedy of all is: Keep the skunk out from under the house in the first place.

ENVIRONMENTAL MEDICINE

Once the investigation has revealed some probable causes, the problems related to IAQ are not over. As with any new field, new labels of expertise are donned by those practicing untested treatment. Dr. Harriet Burge, P.E., while a microbiologist at Harvard School of Public Health, has counseled against the ready acceptance of "clinical ecologists" or "specialists in environmental medicine." Burge observes, "Unfortunately, the methods and theories of these practitioners have not been subjected to rigorous scientific study... The dangers involved in using unproved methods of diagnosis are several." After noting, the patient may not receive the help they actually need or the treatment itself might be deleterious, she concludes, "Finally, the drastic changes in life-style often recommended by environmental medicine practitioners to implement avoidance can be expensive, both financially and emotionally, and probably rarely, if ever, provide a permanent solution to the underlying complaint."

Burge's warning has merit. Every new field has its share of charlatans. These words of caution, however, should not be taken to suggest that any individual or firm self-labeled as clinical ecologists or specialists in environmental medicine are necessarily engaged in some level of quackery. It does remind owners, managers and technical support personnel that the credentials and references of anyone purporting to be an IAQ professional should be checked. Furthermore, the potential employer and manager of such specialists needs to know enough about indoor air pollutants to know the special expertise required to do the job.

THE MANAGEMENT CHALLENGE

IAQ problems are increasingly prevalent, pervasive and even pernicious. Wider application and ongoing revisions to the ASHRAE Ventilation Standard end the challenges of other environmental forces and drivers will force us all to consider the indoor environment even in "healthy" buildings. *Managing* that internal environment has become a major administrative responsibility. Much of the IAQ literature has been written by researchers and consultants for other researchers and consultants. This book is not intended to be an academic textbook or reference, rather it has been developed to provide practical guidance to owners, facility managers and those with facility-related responsibilities. It also offers some special insights regarding management concerns for those IAQ practitioners and professionals who provide technical, legal, remedial, and financial support to the building management team.

Chapter 2

INDOOR AIR QUALITY IN RETROSPECT: HOW DID WE GET HERE?

High school students in an urban institution are sent home ill. Workers in a suburban courthouse strike and then sue because of building induced complaints. Teachers walk out of an urban school because of illness and odors. A hospital emergency room is evacuated because of chemical odors. Public workers enter into a class action suit against the building owner because of permanent disabilities caused by new carpeting. The PTA sues the School Board for injuries to students incurred during an inept asbestos removal. A public library is evacuated and abandoned because of long-term water damage and subsequent mold. A public municipal building is abandoned because of humidity control and moisture problems. University students revolt from taking classes in a lecture building known to cause headaches after just one lecture (regardless of the lecturer). Office workers sue the building owner for illnesses and disabilities caused by exposure to chemical fumes during renovation. A homeowner sues her insurance company for mishandling her water damage claim causing her family adverse health effects from resulting mold infestation and remediation. Tenants in an apartment building vacate their leases and sue the apartment building owner and contractor for problems resulting from leakage through the envelope. Unionized labor in a high-rise hotel in a resort community forced a leading hotel chain to abandon a $96 million tower due to mold infestation in the guest rooms with remediation cost at $72 million. All of these real headline situations illustrate a problem in common: the quality of the indoor air.

Problems with indoor air quality are not new. They undoubtedly reach back to combustion contamination from fires in the caves. A retrospective examination of the many factors leading to our current indoor air quality concerns helps to establish the background for the difficulties posed by internal environments today.

IAQ IN ANTIQUITY

Modern social attitudes toward indoor air quality date back to antiquity. Egyptian historians have found notations of respiratory disease caused by silicate dust from stone cutting as early as 1500 B.C. A thousand years later, Hippocrates described the effects of air on health in his writings. Pliny the Elder, in early Roman times, urged stone masons and asbestos miners to wear masks while working.

The Roman baths were designed to control the volume of water, light, air heat and sound in any given room through room orientation, window placement and the positioning of fountains, plumbing and furnaces. This marriage of the building envelope and its orientation to the heating, ventilation and water distribution systems was a precursor of today's architectural and engineering attempts to provide an acceptable indoor environment.

While concerns regarding comfort and climate did not cease during the early middle ages, they got a fresh impetus late in the period from increased urbanization and the associated pollution problems. Records show that the unbearable stench of the Fleet River caused a Carmelite monastery to petition King Edward I in 1520, for relief from the filth that was interfering with divine services and causing death among the monks.

Large-scale, energy-related indoor air problems can be traced to the burning of coal in the 13th century. By 1273, a law was passed prohibiting the use of coal in London due to the smoke. Parliament petitioned the King to stop the burning of coal altogether in 1306, as the fumes were considered dangerous to health.

Perhaps the most notable incident of this period was the proclamation by King Edward III forbidding the dumping of garbage into the Thames and Fleet rivers when the stench imperiled all London in 1357. Designated as the "Year of the Great Stink," a warm summer brought eutrophic conditions rich in nutrients to the rivers, causing excessive algae growth, bacterial depletion of oxygen, and tons of dead fish.

BODY ODOR, TOBACCO SMOKE, AND DISEASE:
FACTORS OF TRANSITION

Through the years, body odor and tobacco smoke were the prime factors in assessing the perceived quality of the indoor air. Cleanliness has not always been next to Godliness. The decline in the practice of bathing during the Middle Ages, which persisted through the Renaissance, can be attributed to the attitude of the Christians toward cleanliness. The Church attacked the preoccupation with body comfort and attractiveness offered by bathing as tending toward sin. The "odor of sanctity" prevailed and lice were called "pearls of God." St. Paul is said to have observed, "The purity of the body and its garments means the impurity of the soul."

More frequently the lack of personal hygiene during the Renaissance was an economic concern. Poor people worked and slept in the only clothing they owned. While the rich owned more changes of clothes, there is little evidence that they were laundered between wearings. This historical perspective helps explain the current emphasis upon the occupant as the primary pollution source. This has resulted in occupant population driven ventilation rates.

Smoking tobacco has alternately been accepted and rejected by society and the law. King James I was the first to denounce the habit as a "corruption of health and manners." During the 17th century, most of Europe severely penalized or forbade the consumption of tobacco. In 1911, 14 states prohibited cigarettes for moral and/or health reasons. Today these concerns are reflected in the prohibitions governing the sale of tobacco to minors, advertising restrictions, and the increasing limitations on areas where smoking is permitted. Many locales have demonstrated their concerns about the potential health effects of tobacco smoke by severely restricted smoking practices in public access buildings, especially following the aborted OSHA ruling in 1994.

Communicable disease followed a similar historical path as cleanliness and bathing. Yet, knowing nothing about bacteria, virus, and other pathogens, early settlers found that swamps brought disease and conversely, clean mountain air purged the body of fatal "consumption." This led to the development of sanitariums for TB rehabilitation and the use of high ventilation rates in health care facilities that were highly congested and contaminated with communicable pathogens.

These three indoor air quality factors, body odor, communicable disease, and tobacco smoke, or "ETS" (meaning Environmental Tobacco Smoke which refers to the side stream fumes and the exhaled smoke from the active smoker that becomes part of the common breathing air stream), strongly influence our standards today. The need for fresh air has historically been measured in the need to counteract these human-generated pollutants; thus, the common ventilation requirements are for so many cubic feet of air per minute *per occupant*. The sources of many contaminants today; e.g., building materials, combustion, cleaning supplies, etc. unfortunately have very little relationship to the number of occupants in a building.

As we move into more modern times, it becomes clear that indoor air quality has had its own "Back to the Future" scenario. With the development of the first nuclear powered submarine in 1954, submarines suddenly had the capability to remain submerged for weeks (or months) at a time. This required a means of controlling, cleaning and revitiating the quality of indoor air. Through the use of special ventilation and filter systems, air conditioning, chilled water systems, main oxygen supply, CO_2 removal, $CO-H_2$ burners and atmosphere analyzing systems, the internal air in a modern submarine can be maintained at a designated quality level. The designers of the Nautilus' environmental system were ahead of today's building designers by nearly half a century. More recently these advancements have again been illustrated by the space shuttles in which comfortable and safe environments have been designed, installed, and maintained in outer space. Thus, the technology currently exists to create acceptable and safe indoor environments without access to unlimited quantities of theoretically fresh outdoor air.

IAQ TODAY

What has brought about the more recent concern for the quality of air we breathe? Obviously, the IAQ issue did not manifest over night. Many authors and experts focus on the mid-70s with its energy crisis as the spawning ground of today's IAQ problems. Skyrocketing energy costs led to tight building construction that often resulted in drastic reductions in ventilation air and infiltration. This yielded the early and now archaic expression "Tight Building Syndrome." Unfortunately, this label was not only misleading, it was inadequate in providing a full explanation for the issue. It resulted in the focus of blame on ventilation and its inherent energy cost as being the "cause all"—"fix all"—of indoor air quality (and this misconception still prevails in the "Green Building" movement). This is simply not the case, then or now. In fact, increased ventilation in some instances will not help the problem, and could even make matters worse without a fuller understanding of the complex causes of IAQ problems. Let's explore the driving forces that contributed to the maturing development of IAQ.

The Energy Response

When energy prices climbed by more than 600 percent in the 1970s, many well meaning energy managers and engineers designated energy conservation above comfort. Readers are reminded of the "Building Emergency Thermostat Settings: (BETS)" of the Carter era with the accompanying "T-stat police." This "feeding frenzy" of energy savings especially impacted the schools. With their rigidly constrained budgets, schools were among the first to seek Band-Aid approaches to their dilemma. Those "Band-Aids" frequently showed up as duct tape, plastic film, and plywood patches over air intake grills and windows.

Schools were not alone in such responses. Little was known about energy efficiency, but the obvious high cost of conditioning outside air made the reduction in air intake a natural choice—right after adjusting the thermostats. Mechanically induced outside air requirements generally dropped from 15 or more cubic feet per minute (cfm)/occupant to 5 cfm/person and less, even zero mechanically induced outside air.

As we became more sophisticated in energy efficient practices, delayed start-ups and early shut downs to let the systems "coast" became standard operating procedures. Ventilation was designed to rely more heavily on mechanical systems with recycled air. The cracks and leaks were patched and additional insulation was added to walls and ceilings. Variable air volume (VAV) systems became a common modification to provide varying air supply through the distribution system. This was designed to completely shutdown air supplied to zones with low or no occupancy demand for heating or cooling. This allowed energy saving as conditioned dilution air was not delivered to zones during their low demand periods. The result is that air within certain zones became "aged" or stale and stuffy. It also put the system out of balance and imposes part load operating conditions on the air handling and refrigeration system. The more subtle, but more disastrous consequence from an IAQ standpoint, is that contaminants remaining in the space tend to deposit onto and into fleecy and porous surfaces. This created a "sink" effect allowing the pollutants to re-emerge later over time by outgassing due to changes in temperature, humidity, and airflow. The cost-saving, energy efficient building, was born. Unfortunately, we failed to appreciate what a high price tag those cost cutting efforts may carry.

The response to high energy prices carried another cruel twist. As owners sought ways to pay the utility bill, they looked for other places in the facility budget to cut costs. The most elastic part of the budget seemed to be operations and maintenance staff and materials. Those cuts only exacerbated the indoor air problem, for one of the greatest sources of indoor air pollutants is inappropriate or insufficient maintenance. The effects of maintenance cuts were compounded by the energy-efficient tight building which is far less forgiving of poor maintenance.

The Rise of "Least Cost" Design and Construction
While energy management tactics were flourishing, the concept of "value engineering" emerged. This was an engineering concept originally built upon the intent of streamlining, efficiency, and cost effectiveness. Unfortunately, value engineering has lost this once lofty and positive connotation. Now we know that it really means cost cutting, cheapening and "devaluing." This "least cost" mentality lead to many design/construction decisions and tactics that saved on first cost. However, both the owner and the occupants often pay a deferred price in increased life/cycle costs, discomfort, illness and health costs, as well as lowered productivity. In the extreme, first cost savings return their dubious "value" multifold in the form of lost lease income, lost productivity, and even litigation. In the later chapters of this book, we will discuss the concept of "resiliency" engineering, which is a reaction to value engineering and the "least cost" mentality.

Technological Changes
But that wasn't all. Changes in product technology introduced new building materials, products and furnishings into the indoor environment that emitted large numbers of chemicals into the air. Among the culprits were carpet backing and adhesives, wall coverings, paints, stains, paneling and ceiling tile. Polymers and plastic resins became more widely used in construction components as glues, binders, and soiling retardants. Formaldehyde (HCHO) and other aldehydes were typical ubiquitous components for furniture, fabric, plastic surfaces, particleboard and other engineered wood products. There was also a widening usage of fabrics and fleecy materials in the space. Polymeric glues and highly solvent-loaded adhesives were used to attach them. Occupants exposed to such contaminants often complained of irritation, discomfort and other flu-like symptoms.

The effects of many of chemicals emitted from the products are still not fully understood, but many are known or suspected human irritants and some are suspected human carcinogens. Many years may yet go by before the dangers associated with the use of certain chemicals are realized and brought to public attention. Consider, for example, the past and present uses of tobacco, formaldehyde or solvents used in glues. To this chemical stew, we've added a host of cleaning materials. Even such familiar things as floor wax and the emissions from dry-cleaned clothes can increase chemical contaminant levels.

From the old manual typewriter and hectograph gel, we have progressed to computers and photocopiers. What modern office could survive without the high speed copier, the fax, the laser printer, and the desktop computer terminal? In addition to their heat load, these modern conveniences bring ozone, Volatile Organic Compounds (VOCs), and the sub-micron sized respirable particles from the print toner into the modern office environment. If all these technological marvels don't emit harmful contaminants in their operation, there is a good chance the cleaning agents used to service and maintain them will. Frequently, this equipment has been added to old facilities that do not have sufficient air flow to satisfy the newer operational needs. It is not unusual to find copiers stuck off in a corner, hemmed in with files, or set in some unventilated office niche. Because the original HVAC design cannot take into account these activities, irritating or dangerous levels of pollutants can accumulate in pockets of stagnant air.

Building Use Changes

Changes in building use during the life of a building are common. Altered programs, functions, personnel needs, and tenants frequently have necessitated facility modifications. Other changes have been mandated by federal laws, such as handicap access legislation, or through state statutes and local ordinances. New equipment or alterations to the electrical or lighting systems have prompted building changes. Spatial modifications, particularly partitioning and the 1970's "open space" facilities, have changed interior layouts. For example, high density office layout is now possible through "office systems." In addition to the VOC outgassing of their fabric and fiberboard components, work stations can add unpredicted barriers to air flow and ventilation effectiveness. Further, they facilitate increased occupancy density that can easily exceed the original design parameters. Most of the time, corresponding changes in the mechanical or distribution systems have not kept pace. Most every facility manager is aware of pockets of "stale" air, or areas where over (or under) heating and cooling have resulted from such changes.

Another change affecting the building stock is that "Time marches on!." Few design professionals and building managers recall that the OPEC crisis that began the energy panic was in 1973—decades ago. Buildings constructed in the early stages of this period now have aged considerably. Systems, controls, decorations, roofs, elevators, and HVAC systems have been subjected to decades of use and abuse. Poor maintenance, poor filtration, low ventilation rates, water leaks and "least cost" purchasing decisions take their toll. For example, it was revealed in testimony during the trial concerning the EPA Waterside Mall headquarters building in Washington that the ductwork in the building had "only a 1/32 inch accumulation of particulate matter." This equated to *tons* of contamination that had to be cleaned out of the building air distribution system. This had accumulated over years of operating with low efficiency blanket filter media on the outdoor air. Thus, the aging process alone can induce an inevitable incremental degradation in the operating effectiveness and efficiency of a building.

The People Factor

Occupants in this changing internal environment have been bombarded with IAQ bad news in all the pertinent trade publications, most network TV news shows and many hometown newspapers. The general public has been conditioned to environmental issues with the focus on radon, asbestos, PCB, insecticides, fertilizers, lead, and mold. Problem buildings with litigation, like the EPA building and numerous high profile public courthouses, make great headlines. Thus, your tenants and even their families know about air quality in their space. The general public is suspicious of "places that make them sick" and "chemical smells" and "black stuff that comes out of the ductwork." This awareness has raised the expectation of occupants regarding the quality of the air in their personal space. These same people may have experienced physical, emotional and mental reactions. Frequently, the cause-effect relationships are not fully documented. Synergistic relationships between contaminants and ergonomic or organizational stress are not fully understood even today. An individual who is uncomfortable, ill, or under stress is not happy. Further he or she will be more likely to both complain and be more susceptible to indoor environmental concerns.

Those examining IAQ related problems, have pointed to such subtle things as the occupants' inability to control their environment. Though we know that it seldom helps and may even hinder, there was something innately satisfying about having access to and throwing open the window. Job stress and job satisfaction have been suspected as contributors to indoor air complaints; however, Hedge's research in the United Kingdom found no significant relationship. Others have suggested job stress, dissatisfaction or even problems with glare, lighting, or uncomfortable furniture may have prompted increased use of alcohol or drugs. These substances may constitute intervening variables in any cause-and-effect studies related in indoor air quality. A confounding factor to these human relation issues is the too prevalent response by facility managers to IAQ complaints. The Building Operators Magazine reported that many early responses to this issue have been "frustration mixed with skepticism" and a "tendency to downplay complaints." Handling of IAQ problems has therefore frequently been slow or non-existent. Thus, the real problems are complicated by the perception of lack of concern, lack of response, lack of communications and simply poor human relations. This inappropriate response adds anger to the health effects equation and this usually equals litigation, not mitigation.

Several studies have observed that women seem to be more susceptible to some indoor air pollutants and are more apt to complain than men. The quality of indoor air in a given facility is not uniform. Clerical workers, who are predominantly women, are often crowded together and they are frequently in interior air supply zones. The perimeter offices generally receive more air supply and more outside air than interior areas. The interior zones are more likely to stay near the warmer end of the design and thermal operating range. Airborne chemicals from the clerical equipment, such as copiers, increase the clerical staff's exposure levels.

The potential relationship of all these factors help to illustrate the problems researchers have in trying to establish causal relationships and linkages. Likewise, investigators have difficulty in trying to pin down certain pollutants as factors in building related illnesses. Indoor air quality research is reminiscent of the admonition, "If you think you have a solution, you probably don't understand the problem."

The Changing Scene
Federal actions regarding IAQ, especially with reference to asbes-

tos and radon, were initially directed primarily at schools and homes. This movement transferred to the commercial sector as could be anticipated. Commercial facility owners have limited concerns related to radon, but asbestos removal from buildings constructed prior to 1972 have turned budgets inside out. The early focus of the regulations was mitigation through removal. This created tremendous financial burden on both the private and public building management segment. Of course, the regulations and their interpretation have since softened and emphasized other techniques, like control by isolation through containment. The impact of the early policy was to financially burden facility and plant management teams. The unfortunate legacy of these policies has sensitized decision makers who are now even more suspicious of the reality of environmental issues.

This retrospect also serves to remind us: Amenities are not created equal. Nor should they be. Johnson and Curley list the lobby, elevator, heating system and the bathrooms as, "The basic things that are most important to us as 'spec' office builders." These are emphasized more than other facets because of their importance to prospective tenants.

Focus is gradually shifting from buildings that are self-conscious art forms that may attract tenants to the comfort that will keep them. Mikulina has observed, "We must help builders and developers understand that, while an architect can attract tenants through aesthetics, only a good comfort system will retain tenants through increased productivity." A similar architectural trend gaining momentum in the latter 90s and early 21st century is the "Green Building" movement. This trend recognizes more fully the role of the building as an ecosystem or "habitat" and its impact upon both the occupants within and the global environment without. The U.S. Green Building Council has defined the green building to mean "structures that are designed, renovated, constructed, operated, and eventually demolished in an environmentally and energy efficient manner with least impact upon our global and internal environment." Of course, the greening trend recognizes and incorporates the importance of indoor air quality. USGBC has further promoted green construction practices through the development of a certification program labeled LEED. This is referred to as a Green Building Rating System that defines a green building tactics using a common standard of measurement and promotes the use of green design practices through a certification attainment and labeling system that is gaining momentum with progressive owners. As evidence of the recognition of this trend,

ASHRAE, in conjunction with USGBC and IES, has developed and published a new standard on sustainability, ANSI/ASHRAE Standard 189.1-2009 Standard for the Design of High-Performance Green Buildings. There is mounting concern, however, that the LEED building rating system (which is based on modeling) fails to assure improved energy performance in actual building operation and litigation has been instigated against USGBC based upon this allegation.

Unfortunately, this trend also has its dark side. First, sustainability focuses on energy reduction and runs the risk of over emphasizing that factor to the detriment of IAQ similar to the early days of IAQ and "Tight Building Syndrome". Secondly, perhaps as a reaction to the former, there is an inappropriate heavy reliance on ventilation alone that is evidenced by the recommendation by Standard 189 to reward the over-ventilation of buildings to address IAQ acceptability. Lastly, by its nature, green buildings promulgate the wide usage of recycled products—products that may be green and good from one perspective—but bad for IAQ when considered from a contaminant perspective. One such example is the usage of recycled cellulose newsprint as an insulation material. One product is applied wet and loads the structure with water that cannot dry out before fungal infestation can occur. Further, it is an open invitation for other problems such as corrosion of plumbing and electrical systems from the chemical additives. In Southern locations, the termite is also a likely and unwelcome resident.

HVAC as a Source

The building's own HVAC system may be the culprit. The HVAC distribution system is the air pathway (in a sense the lungs of the building) throughout the building. Air pollutants may be moved by the distribution system from an area of the building used for a specialized purpose, such as industrial shops or laboratories, into other areas, such as offices. Polluted outside air, no longer referred to as "fresh air," may be brought in from loading docks, garages, or picked up from a neighbor's or the building's own exhaust. Polluted air, such as automobile exhaust in a parking garage, may find its way up elevator shafts and stairwells into office spaces. Humidifiers, dehumidifiers, air conditioners, cooling towers and ductwork may be the source of biological contaminants spread by the ventilation system.

These conditions can occur or develop because the HVAC system is inherently vulnerable to maintenance practices. The traditional philosophy of "if it ain't broke, don't fix it" matched with least cost construction and purchasing practices spell disaster for even the finest of building designs. Most experts agree that buildings that do not work properly due to poor upkeep are the most likely candidates for Sick Building Syndrome. EPA researchers claimed over half of the complaint buildings they visit are plagued with poor HVAC maintenance.

An additional component of the maintenance protocol is the role of filtration. Low efficiency disposable furnace filters or blanket media were and are still widely used in commercial buildings. While they represent the least first cost, they also represent the least performance in providing contaminant control and protection of the systems, the building, and the occupants. The resulting poor filtration effectiveness allowed particulate accumulation on cooling and heat exchange coils, ductwork, and the facility. This robs heat exchange efficiency and wastes energy because of operating with fouled systems, such as dirty coils. Poor filtration also allows a buildup of nutrients for microbial growth, as well as a source of odor and VOC sinks. Because of the emerging importance of filtration in IAQ assurance, this book now contains a special section on air filtration and provides the reader with practical information about the role of the HVAC systems in buildings having special security and environmental protection needs.

SUMMARY: WHAT DOES ALL THIS MEAN?

Air quality has been a concern for 4,500 years, so it is not really new. As we have learned in this chapter, however, most of the serious focus and development of IAQ has been since the early days of the more modern energy panic, yet it is not just related to energy. The primary message to the reader is that the IAQ issue is broader and more complex than merely an energy or ventilation issue. As this chapter discusses, there are number of trends and modern developments that have contributed to the diversity and intensity of the problem. In the harsh realities of today, the specter of lawsuits, lost productivity and frequent tenant turnover have garnered more and more attention from owners, facility managers, architects and engineers. It is becoming increasingly clear that it is bad business not "to communicate the important message" that a comfortable healthy environment is a productive environment. It is the goal of this book to assist the building owner and manager of the 21st century address indoor environmental needs, but more importantly, to prevent IAQ related incidents from occurring in the first place.

Chapter 3

CLASSIFYING INDOOR AIR PROBLEMS: WHAT CAN GO WRONG?

Owners and managers seldom have the luxury of knowing the full range of constituents of the air in their buildings. If they have an indoor air problem, they know about it because they hear the complaints, symptoms, other medical diagnosis. But they aren't handed a list of potential contaminants. It is important for the reader to know that there are literally thousands of trace components or "constituents" of indoor air. Most of these are either benign or are normally present in concentrations that are below odor perception or below thresholds of concentration that can cause human response, such as discomfort, irritation, or health effects. When these normally occurring levels are amplified or when known harmful constituents are present in harmful concentrations or combinations, it is appropriate to refer to them as "pollutants" or "contaminants." These terms are used interchangeably in the literature and this text. Either term implies a constituent of the space that exhibits either or both harmful properties or harmful concentrations.

Unfortunately, most of the IAQ literature is based on specific and known air constituents and their contamination level. That's not particularly useful to the practitioner who is confronted with an indoor air problem that needs attention *now* with little or no idea what pollutant is causing the problem.

The answer would seem to be: figure out what constituents and/or potential pollutants are causing the problem. But it's not that easy. Identifying specific culprits often proves to be difficult, costly, even impossible to align with specific complaints. The situation can be further confused by the multifactorial nature of indoor air problems. Even if a contaminant in the area has been identified, another undetected contaminant may be the one at fault. Or, they may be acting synergistically to cause or worsen the problem.

Working from the manager's perspective, symptomatology and diagnosis offers a preliminary classification opportunity. It offers a way to get an early handle on the situation from what is already known. This chapter, therefore, looks at the situation as it is first made known to management. We start with the complaints, the symptoms; only then, do we look at major contaminants and environmental conditions to explore their relationship to these health effects. And the very last thing we do is look at air testing and diagnostics.

SICK BUILDING SYNDROME
AND BUILDING RELATED ILLNESS

Sick Building Syndrome (SBS) is a worldwide and complex problem. SBS initially was used to describe a building where a set of varied symptoms were experienced predominantly by people working in an air conditioned environment. Subsequent early study by Finnegan and others has shown that SBS is not limited to air conditioned facilities and can, in fact, be observed in naturally ventilated buildings. More recent joint research in England by British architect Alexi Marmot and her husband Michael, an epidemiology professor, concludes complex SBS involvement with the occupant's home environment, poor facility management, and worker satisfaction.

A syndrome, by definition, is a group of signs and symptoms that occur together and characterize a particular abnormality. Frequently they form an identifiable pattern. This makes diagnosis by exclusion possible. For example, organic lesions are not associated with SBS. If an occupant has persistent organic lesions, it can be assumed that the cause of the sickness is not SBS.

As defined in Chapter 1, a building is said to be "sick," when 20 percent or more of the occupants voluntarily complain of discomfort symptoms for periods exceeding two weeks, and affected occupants obtain rapid relief away from the building. The 20 percent figure is an arbitrary level derived from earlier ASHRAE efforts to define comfort. This acceptability level, when transferred to the IAQ arena as a level of temperature acceptability, may mislead managers who look to "20 percent" as a guideline for action. A knowledgeable investigator would ask, "If you have 3,000 employees and only 10 percent are ill, do you wait? That's 300 people who could be suffering needlessly from a pollutant in the work place."

A problem closely associated to SBS is building related illness (BRI). Medical diagnosis can identify specific health effects that result in known disease etiology, such as Legionnaire's disease, that are a direct result of building conditions. Once diagnosed, a BRI can help identify the source contaminants and may reveal ways to remedy the situation. A BRI facility has almost always passed through the SBS stage and usually still has other contributing contaminants or causes at work.

The five symptom complexes associated with SBS are discussed below; followed by the major building related illnesses.

SBS SYMPTOMATOLOGY

The five SBS symptom complexes can occur singly or in combination with each other. Symptoms may be cyclical or episodic. They may be nonspecific and often resemble a common cold or other respiratory illnesses. They usually get worse as the day progresses, may worsen through the work week, and ease or disappear once the occupant is away from the building for a time.

1. Eye Irritation

A burning, dry, gritty sensation is experienced in the eyes without any evidence of inflammation. Severity varies from day to day. Sensitivity is greater for occupants wearing contact lenses.

2. Nasal Manifestations

The most frequently cited nasal symptom is "stuffiness," which develops rapidly when an individual enters the building, persists while in the building, and goes away quickly upon departure. For some people, this "stuffiness" also is a specific reaction to high temperatures. Other nasal symptoms, which are more variable and apt to be less persistent, are nasal irritation and rhinorrhea. Symptoms are frequently suggestive of an allergic cause.

3. Throat and Lower Respiratory Tract Symptoms

A persistent dryness of the throat, which seldom shows any inflammation, is a principle symptom. The occupant may gain some relief by drinking large amounts of water.

A typical indication of lower respiratory tract difficulty is a short-ness of breath, a sense of not being able to breath deeply, which is not related to any lung infection or bronchial asthma. It is generally relieved by stepping outside to take a few deep breaths.

4. Headaches, Fatigue, General Malaise
The headaches are usually frontal in position, occur in the after-noon and may occur daily. Headaches may range from moderate to severe migraine.
Headaches and related fatigue, dizziness, difficulty in concentra-tion and general malaise are the most frequently cited sick building symptoms.

5. Skin Problems
Dry skin is a frequent SBS complaint, particularly from female occupants. It is considered a building associated symptom when it improves during protracted absences from the facility. Warm dry air or excessive air movement may create a particular type of dermatitis on exposed skin surfaces. Skin rashes or irritation may result from exposure to some contaminants.
SBS may aggravate existing health problems and diseases, such as sinusitis and eczema, but these are considered outside the general SBS symptom complexes.

BUILDING RELATED ILLNESS DIAGNOSIS

Building associated sicknesses, other than those related to SBS, are generally allergic reactions or infections. The allergies include asthma, humidifier fever and hypersensitivity pneumonitis. Bacteria, fungus and virus can cause the BRI infections. While BRIs have some characteristics in common, there are others, as noted below, that help distinguish one from another. Careful documentation of symptoms and complaints as well as their patterns, and predisposing conditions can be used as tools to focus further diagnostic activity.

Allergies
Asthma, Rhinitis
Asthma (which literally means "panting") is a respiratory dis-

ease in which spasm and constriction of the bronchial passages and swelling of the mucous lining cause obstruction of breathing. It is often due to allergy particularly to dust, animal fur/feathers, molds and pollen. Asthma is also a frequently referenced condition resulting from poor indoor environmental circumstances. It is reportedly the number one cause of overnight hospital stays by school-aged children accounting for over one million sick days a years from school. According to the American Lung Association (ALA), asthma incidence has doubled over the last twenty years, primarily in the Western developed countries in larger highly populated urban locales. According to the U.S. department of Health and Humans Services, the incidence of asthma among school-age children increased 160 percent between 1980 and 1994. In response to this alarming increase, EPA has focused on asthma among children particularly those living in urban, low-income communities. As much as 15% of asthma reactions are blamed upon the workplace as the point of origin. The incidence of asthma in children may also double in the presence of ETS in their home. A contributor to asthma symptoms is the condition of "atopy," which genetically predetermines certain individuals to be more susceptible to asthma and allergic rhinitis. The frequency and severity of asthma symptoms varies markedly from one person to another. Attacks may recur in hours or days, or may be absent for months and even years. Symptoms are those generally associated with asthma and increase in intensity with prolonged exposure; i.e., a work week, but improve away from the facility.

Allergic responses of the upper and lower respiratory tracts generally are secondary to the inhalation of allergens. These bring on an overreaction of the immune system that floods the system with histamines, resulting in the allergic symptoms. In younger children, allergen exposure has been shown to increase the risk for earlier onset of asthma. These allergens are usually associated with poorly maintained buildings and may originate in humidifiers, particularly cold water spray humidifiers, contaminated by microorganisms. However, they may also include external environmental pollutants, pollen, respirable particulates, dust mites, pesticides, cockroaches, and environmental tobacco smoke. Many people (over 2% of the general population) are specifically cat allergen sensitive (FEL d 1). Because this allergen is small and very "sticky" it can easily transport from the home into the work place or school via pet owners.

Hypersensitivity Pneumonitis.

This generic term for a common manifestation, is also referred to as "extrinsic allergic alveolitis." It is basically an allergic lung disorder and it occurs in varying intensity. After a week to ten days, symptoms usually regress without further exposure. Symptoms get increasingly worse during exposure and show some relief when causal agent is absent, such as a weekend. Malaise and myalgia (muscular pain) are almost always present. Headaches frequently occur. In the more acute phase, shortness of breath, fever, chills and a dry cough appear within an incubation period of about six hours. Attack rate is very low. Individual susceptibility is suspected as a key factor.

Humidifier Fever

Symptoms of humidifier fever are "flu-like," lethargy, arthralgia (neuralgic pain in the joints), myalgia and fever. Sometimes headaches, polyuria (excessive urination) and weight loss occur. More severe symptoms include shortage of breath and coughing. These systemic and respiratory symptoms occur with initial exposure; i.e., the first day of the work week. They progressively improve during exposure and in the absence of exposure, only to recur on re-exposure. In a work setting, symptoms appear on Monday, improve during the week and weekend and recur the following Monday. This pattern clearly distinguishes humidifier fever from hypersensitivity pneumonitis.

During the height of the reaction, medical examinations reveal the presence of late inspiratory crackles during auscultation ("listening" to the chest) and impaired gas transfer in the lungs. The chest radiograph is always normal and the lung function is normal between attacks.

The cause or causes of humidifier fever are not known. Some organisms have been isolated and incriminated during outbreaks, but as yet not indisputably identified. However, immunological investigations almost always reveal the presence of precipitating antibodies to antigens extracted from the humidifiers. In bronchial provocation tests, water from the humidifier usually reproduces the symptoms and physiological changes.

Attack rates vary considerably in reported outbreaks. Age appears to be a factor and appears to be associated with the duration of exposure and the development of antibodies. Highest incidence rate is in the winter (probably due to seasonal humidifying activity).

In addition to the tolerance pattern, or the "Monday Morning

Phenomenon" of humidifier fever, it is distinguished from hypersensitive pneumonitis in other ways. Humidifier fever shows no decrease in lung function or pulmonary fibrosis, there are no radiographic changes and it seems to be brought on by comparatively low levels of antigen whereas hypersensitivity pneumonitis is associated with massive antigen exposure.

Infections

Bacterial

LEGIONNAIRES' DISEASE, caused by *Legionella pneumophila*, is the most widely recognized bacterial form of BRI infectious disease. Features typical of Legionnaires' diseases are headache, chest pain, vomiting, diarrhea, weight loss, fever, dry cough, recurrent chills, dyspnea, myalgia, abdominal pain and pneumonia. It has a 15 percent fatality rate.

Legionnaires' disease is not a recent phenomena and has been backdated to 1947. While occasionally appearing in epidemic proportions in certain environments, it is more apt to be sporadic in nature. Occurrence has been estimated as high as 116,000 cases per year. Its reemergence has been tied to an outbreak during a convention of the American Legion in Philadelphia in the mid 1970's (thus, the coined name Legionnaires' Disease or Legionellosis). According to Don Millar, previous head of NIOSH, there may be as many as 8,000 fatalities annually due to Legionella derived pneumonia. This number may be even higher as health officials at CDC estimate from 25,000 to 100,000 cases annually with 5 to 15% of those reported being fatal. Many more persons may be infected with the Legionella bacterium. However, definitive reporting is difficult because the disease symptoms are either mild or so similar to other pneumonia forms that mis-diagnoses occur. Recent data from CDC indicate that Legionnaires disease is on the rise in the US, which may mean more outbreaks, or it could mean that more cases are being accurately diagnosed and reported.

The causative bacteria are water bound and occur anywhere water is allowed to age, thereby enabling the bacteria to amplify. Thus, outbreaks have been tied back to contaminated decorative water falls, jacuzzi or water therapy treatment, cooling tower water (in the case of the Legionnaires' convention), condensate pans, showers, vegetable misting equipment, and even potable drinking water systems. Because the bacillus must be airborne and survive to be ingested into the deep lungs, the primary route of transfer is in the form of fine mist or aerosol.

Thus, the source must both be aged water with a high concentration of bacteria, but also must be a source of fine aerosol spray, such as a cooling tower plume. One is reminded of ancient but still valid advice from Hippocrates in the 4th century "Do not drink from stagnant water."

The incubation period for Legionnaires' disease at onset is two to ten days and the attack rate is a low 6 percent. It generally favors males and individuals over 55. Smokers are roughly 2 to 5 times more likely to get the disease. It appears to have a summer-fall seasonality.

Building managers need to be aware of this disease risk particularly with regard to workers involved in the maintenance of dormant or stand-by cooling towers which have not been actively treated or monitored. ASHRAE has developed a proposed standard BSR/ASHRAE 188P-201x Prevention of Legionellosis Associated with Building Water Systems, which is now in the public review and in the final publication stage. The Standard show how to reduce the likelihood of Legionella transmission by identifying the critical control point in a building's water system. It is advisable for any water system that will be subject to aerosolizing, such as waterfalls or cooling towers to be monitored through a proactive water testing program to avoid the "negligence" allegation should an outbreak occur

Pontiac Fever

This is a relatively mild clinical form of Legionnaires' disease. Often related to specific worker activity it can occur with a high attack rate from a common source, such as cleaning steam turbine engines. Fever, malaise, headache, chills, myalgia, nausea and diarrhea are also features of Pontiac fever. Respiratory symptoms, such as sore throat and a slight cough, may be present, but pneumonia is *not* associated with the disease. This nonpneumonic form of Legionnaires' disease has a shorter incubation period, 5 to 66 hours, and an almost 100 percent attack rate. No age or sex distinction is evident. Outbreaks occur in summer. No fatalities have been reported. The milder form of the disease is thought to be brought on by residual non-viable components of the Legionella bacteria, such as chemical endotoxins that remain in inhaled aerosol.

Medical diagnosis of Legionnaires' disease and Pontiac fever can take some time, as the necessary serologic diagnosis requires the appearance of antibodies.

Tuberculosis

Coined as "TB," this infectious disease is caused by the bacillus *Mycobacterium tuberculosis* that is transmitted through sputum, either in airborne droplets or by dust particles. It was discovered in the late 1800's and has been under control through modern antibiotics. According to Nardell, TB is reemerging as the single most lethal infection in the world having the greatest morbidity and mortality of any infectious disease. It was projected by WHO that up to 300 million would be infected over the next decade, of which 30 million will die globally. The current pace of the disease is approximately .4 percent per year but much faster in Africa and countries of the former Soviet Union. Although found mostly in third-world nations, the disease is most treacherous in health care settings where a multiple drug resistant (MDR) strain is experiencing nosocomial transmission. Although cases are subsiding, this is also true in the United States, especially in the congested inner cities of our major urban centers where homeless centers and other high population densities, such as prisons, facilitate the airborne transmission of the tuberculosis bacillus. Individuals having compromised immune systems are particularly vulnerable. Because of the special needs of the health care facility and the unique aggressiveness of this communicable disease, the Centers for Disease Control and Prevention (CDC) issued definitive mechanical and operational guidelines for the control and prevention of TB. This protocol included administrative measures, such as work practices, education and training, and TB screening of health care workers. Further, additional tiers of the recommendations include engineering measures such as filtration, pressure barrier relationships and UVGI, as well as personal protection for the health care worker. These measures, in large part, are responsible for declining cases of TB in the US unlike the rest of the world. The CDC has incorporated these practices in their 2005 publication (*Guidelines for Preventing the Transmission of Mycobacterium tuberculosis in Health Care Facilities*). Included in the document are recommendations regarding the usage of UVGI (ultraviolet germicidal irradiation) in the occupied space.

Fungal and Viral Infections

The list is long, as over 100,000 fungi worldwide have been described and up to 2,000 are added each year. Infections, for example, have been caused by the fungus, *Aspergillus* (Aspergillosis), coming into the buildings from outside air and contamination in the duct work has been well documented. Of particular concern are old or immunocompromised patients in hospitals. Some fungi produce volatile organic compounds; so VOCs can be of biological origin (MVOCs or microbial volatile organic compounds).

Other examples of infections that are more rare and exceptional but still threatening include the following: Histoplasmosis which is a disease with flu-like symptoms caused by a fungus carried in bird droppings or bat droppings; and *Psittacosis*, also called parrot disease, that is carried by birds, like pigeons and transmitted by dust particles which carry the causative parasitic bacterium.

With the greater attention indoor air quality is receiving, it is easy to overlook the viral infections. This area has been well treated in the medical literature and is not addressed here. It is interesting to note, however, that an air-conditioning system appeared to be the means of spreading an epidemic infection of measles in a school.

Health Symptoms Summary

Table 3-1 summarizes health symptoms and associated SBS and BRI illnesses. The table is particularly useful in identifying which illness is the only one to exhibit a certain symptom, or only illness that does not exhibit that symptom. This helps with the sorting process. For example, eye irritation is only identified with SBS; not BRIs. Humidifier fever is the only illness to exhibit lethargy but not general malaise. *Malaise* is a vague feeling of uneasiness or physical discomfort. *Lethargy* is characterized by abnormal drowsiness or torpor, apathy, sluggishness and great lack of energy. When a building associated illness is suspected, then, abnormal drowsiness without any feelings of physical uneasiness would suggest humidifier fever. If these symptoms are accompanied by the "Monday Morning Phenomena," the possibility of humidifier fever is even greater.

Table 3-2 offers some key factors that help distinguish building related illnesses (BRI) from each other. Certain factors, such as seasonality, help identify the illness. Other information, such as predisposing factors, suggest situations where preventive action may be more critical.

It is important to recognize that many of the SBS and BRI symptoms described above may *not* be caused by the building. A sudden outbreak of flu-like symptoms may, in fact, be the flu and not a building-involved exposure or cause. Thus, related but coincidental health complaints may be confounders to accurate conclusions.

Table 3-1. Health Symptoms and Associated SBS and BRI Illnesses

ILLNESS	Chest Pain	Chills	Concentration Diffi.	Cough	Dizziness	Eye Irritation	Fatigue	Fever	Headache	Lethargy	Malaise	Muscle Ache	Nausea	Pain in Joints	Pneumonia	Shortness of Breath	Skin Irritation	Weight Loss	Other
SBS			■		•	■	•	•	•		•					•	■		Dryness of throat nasal stuffiness rhinorrhea
Legionnaire's Disease	•	•		•				•	•			•			■	•		•	Abdominal pain, confusion, diarrhea, vomiting
Humidifier Fever	•	•		•				•	•	■		•		■		•		•	Polyuria
Hypersensitivity Pneumonitis		•		•			•	•	•		•					•			
Pontiac Fever	•	•		•	•			•	•		•	•	■						Diarrhea, sore throat

Table 3-2. Building-related Illnesses*

BUILDING RELATED ILLNESS	ATTACK RATE	INCUBATION PERIOD	SEASONALITY	PREDISPOSING FACTORS	RELATED COMMENTS
Hypersensitivity Pneumonitis	3-16%	6 hours	—	No apparent disposing factors	Only small proportion of exposed develop active disease
Humidifier Fever	Variable 2.6 to 70%	4 hours	winter	SM: negative correlation between smoking and presence of antibodies Ex: duration of exposure, development of antibodies and age	"Monday morning phenomenon"
Legionnaires' Disease (LD)	6%	2-10 days	summer - fall	Age: Above 55 years Sex: Males SM: more common in smokers	Avg. 15% fatality rate Pulmonary involvement More common in those with underlying disease
Pontiac Fever	95-100%	5-66 hours	summer	Age: Lower than LD but reflects age distribution of exposed Sex: reflects sex distribution of exposed (but high attack rates)	No fatalities Lack of pulmonary involvement

*Adapted from work by J.M. Benard, Montreal, Canada

PUTTING HEALTH INFORMATION TO WORK

Because epidemiologists, engineers, architects, physicians, industrial hygienists, researchers and regulators all have their own approach to addressing indoor air quality, each group has emphasized different aspects and the relative importance of various diseases. One of the great weaknesses remaining in indoor air quality research and problem resolution available to practitioners is the absence of cross discussion and consensus among the disciplines involved.

Others have grouped building associated maladies in order to define sick buildings, or the rate or prevalence of certain symptom patterns. Early work by Robertson *et al.* grouped the symptoms as dry symptoms, allergic symptoms, asthma symptoms, and uncertain cause. Hodgson and Kreiss using an epidemiological approach, developed the following groupings; allergic respiratory disease, mucous membrane irritation, infections, dermatitis, reproductive complaints, miscellaneous, and tight-building syndrome.

In an effort to relate symptom clusters to the type of building ventilation, Hedge *et al.* established three general factors; general health, mucous membrane, and musculature. In their work, they found no "mucous membrane" symptom clusters in naturally ventilated buildings. The results of their research also revealed that symptoms, such as skin dryness, fever and respiratory problems, did not form significant clusters even within air-conditioned buildings.

Clearly no hard and fast grouping of symptoms can unequivocally establish a medical paradigm. The paradigm used in this chapter most closely follows the system complexes associated with SBS and the specific BRIs used by the Joint Research Centre—Institute for the Environment of the Commission of the European Communities.

It should be stressed that most of the research reported in the literature, upon which this examination is based, does not meet stringent scientific criteria. Information on the investigation of healthy buildings, for example, was and still is woefully lacking. Few studies have used "control" buildings and investigators did not follow an established common protocol. When used, efforts to establish controls as representative of the population of control buildings has not been clearly documented.

In whatever fashion complaints and symptoms are grouped, attributing them to contaminants and building sources, wherever possible, is of critical importance. Table 3-6, at the end of this chapter, lists the symptom complexes and diseases with probable causes by contaminants, environmental conditions and primary sources. To make full use of the table, some knowledge of the probable causes, both contaminants and conditions, is important.

CONTAMINANTS AND THEIR SOURCES

Contaminants may originate outside the building to be transported or ventilated into the facility, or they may be generated in the building. While in the building, they have a limited number of options available to them; The indoor pollutant flow shown in Figure 3-1 provides an overview of the contaminant's "life" in a building.

Indoor air quality is determined by a range of conditions and the interactions of "sources," "sinks" and air movement among rooms and between the building and outside. "Sources" as depicted in Figure 3-1 may be building materials, furnishings or the HVAC system. Other sources are consumer products, office equipment, and purposeful activities, such as pest control. The occupants themselves constitute a major source of pollution, especially smokers. "Sinks" are high surface area or porous sites that odors or other gaseous contaminants deposit upon or within. They may be located in the rooms or systems and may ultimately become secondary sources themselves. Air movement in a building consists of (a) natural air movement among rooms sometimes fostered by occupant movement, (b) air movement driven by a forced air system (HVAC); air movement between the building and outside through ventilation, infiltration and exfiltration; and (c) air movement driven by elevator piston action, thermal stack effect, and air pressurization differentials.

As noted earlier, most discussions of indoor air quality are discussed from the perspective of specific pollutants. Usually an attempt to organize the pollutants in some fashion is made. As with symptoms, the organizational patterns seem to be as diverse as the disciplines doing the work. Early efforts by Woods sorted pollutants by physical characteristics, "mass" and "energy." A few have sought to address contaminants by their source; e.g., those entrained in a building, combustion products, activities of humans.

The composition of chemicals known collectively as indoor air pol-

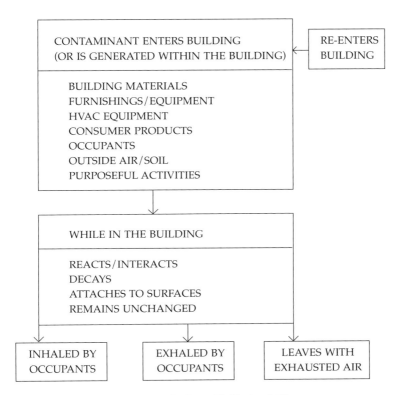

Figure 3-1. Indoor Pollutant Flow

lutants is a complex array of constituents made up of gases, vapors and particles. Determining health effects related to these pollutants collectively, individually, or in certain combinations requires extensive information about the exposure of an individual to this mixture. A body of literature is developing that provides information in varying degrees on the major indoor pollutants affecting human health. These pollutants can be divided into particles (solids and liquid droplets) and vapors and gases.

Particles of specific interest include:

1. Respirable particulates as a group (10 microns or less in size);

2. Tobacco smoke—solid and liquid droplets—as well as many vapors, odors, and gases;

3. Asbestos fibers;

4.   Allergens (pollen, fungi, mold spores, animal dander, insect parts and feces);

5.   Pathogens (bacteria and viruses), almost always contained in, or on, other particulate matter (also referred to as "viable particles").

Vapor and gases of particular interest include:

1.   Carbon monoxide (CO);
2.   Radon (decayed material becomes attached to solids);
3.   Formaldehyde (HCHO);
4.   Other volatile organic compounds (VOCs);
5.   Nitrogen oxides (NO and NO_2); and
6.   Odors.

Increasingly, the procedure has been to treat the major indoor pollutants, either individually, or in small groupings; e.g., bioaerosols, which includes allergens and pathogens.

Since the practitioner is most apt to be concerned about the relationship of particular health manifestations to particular contaminants or practical groupings, the latter approach is employed here. In this way, the facility manager, who is wondering if wet carpeting in the lounge could cause asthmatic reactions can turn to bioaerosols. Or, the plant engineer, who may be trying to discern a relationship between the HVAC leaks he has just detected and coughing and wheezing among occupants, will find the approach workable. Certainly those in a facility responsible for providing a safe, comfortable, productive environment worry about tobacco smoke as an entity and not about the more than 4,700 chemical compounds that come from a lighted cigarette.

The reader should be aware that none of the classifications, including the one used here, is totally satisfactory, as there are considerable overlaps and co-mingling within and between the classes of contaminants. Tobacco smoke, as just indicated, is an obvious example. The joint effect of sidestream and mainstream tobacco smoke to the passive smoker is an exposure to both particulates and gases, including a long list of volatile organic compounds. The organization of contaminants on the following pages then, was predicated on the most useful treatment for the practitioner, but many of the pollutants as listed are *not* necessarily mutually exclusive, nor do they occur singularly. Not only do airborne contaminants comingle and confound their impact on occupants, but other environmental factors, such as lighting and thermal conditions, influence the perception of air quality. These factors are discussed in a newly published document from ASHRAE entitled Guideline 10 Interactions Affecting the Achievement of Acceptable Indoor Environments.

In the following discussion, the description, the probable sources, the symptoms and health effects are discussed for each major contaminant category. Later chapters discuss control procedures as well as laws/regulations/standards.

ASBESTOS

Asbestos is a term used to refer to a number of inorganic minerals that have specific properties in common. The serpentine mineral, chrysotile, is the most commonly used asbestos and represents about 95 percent of the asbestos used in buildings in the United States. The second largest asbestos group is amphiboles, which includes amosite and crocidolite. The fiber structure and the associated health risks are very different by type. Once in place, asbestos does not degenerate spontaneously. Fiber bundles are not inclined to be disrupted without some mechanical external disturbing force.

Sources
Man has used asbestos for thousands of years. Until the 1970s, it was the material of choice for thermal and acoustical insulation as well as fireproofing and flame retardency. It can be found in thousands of commercial products including reinforced cement, older vinyl tile, siding and roofing, and heat-resistant textiles.

In the building structure, asbestos containing material (ACM) is most frequently found in boiler insulation, pipe insulation, sprayed-on fire proofing, breaching insulation, floor and ceiling tiles. The EPA provided in early estimates that 20 percent, or 733,000, buildings in the country contained friable asbestos—and this figure excluded schools and residential buildings with less than ten units.

Symptoms and Health Effects
There are no immediate discernible symptoms of asbestos exposure. EPA and medical specialists have estimated that 1,000 to 7,000 al-

ready exposed people will die of asbestos-related diseases over the next few decades.

All current information on the health effects resulting from asbestos inhalation come from studies of occupational settings with high exposure levels. Most data are based on the amphiboles group, not common in U.S. construction. Three forms of disease have been associated with the inhalation of asbestos fibers: (1) asbestos-fibrosis or scarring of the lungs; (2) mesothelioma—a malignancy of the linings of the lung and abdomen; and (3) lung cancer.

Asbestosis—deaths have only been observed among individuals that have been occupationally exposed to high levels of asbestos. There does not appear to be any evidence that asbestosis should be a concern as a result of environmental exposures today.

Mesothelioma and lung cancer—conservative assessments by various researchers, as indicated in the Harvard report, "Symposium of Health Aspects of Exposure to Asbestos in Buildings," place the associated lifetime risk of death at 1 per 100,000 for 10 years of occupancy in buildings with 0.001 regulatory fibers/ml of mixed fiber types. Recent data suggest average level of asbestos in schools and other buildings with ACM is generally well below 0.001 fibers/ml. The Harvard researchers observed that the lifetime risk of one premature death per 100,000 associated with asbestos exposures was small compared with other environmental risks. The data in Table 3-3, taken from the Harvard report, offer some comparative risks by cause and affirms the low relative health risk of asbestos exposure.

The primary health risk among occupants would appear to be to operations and maintenance people who, in the course of their duties, may disturb ACM. The greatest health risks are, of course, to those who remove the material. OSHA guidelines for asbestos removal should be followed to minimize this risk.

Mounting evidence that asbestos is not as great a problem as previously surmised prompted Congress to request the EPA through the Health Effects Institute to reassess asbestos-related health risks.

BIOAEROSOLS/MOLDS

Bioaerosols, or airborne biological agents, include fungi (yeasts and molds), dander, spores, pollen, insect parts and feces, bacteria and

Table 3-3. Published Estimates of Risk from Various Causes (Mainly U.S. Data)

CAUSE	LIFETIME RISK OF PRE-MATURE DEATH (PER 100,000)
Smoking (all causes)	21,900
Smoking (cancer only)	8,800
Motor Vehicle	1,600
Frequent Airline Passenger	730
Coal Mining Accidents	441
Indoor Radon	400
Motor Vehicle—Pedestrian	290
Environmental Tobacco Smoke/Living with a Smoker	200
Diagnostic X-rays	75
Cycling Deaths	75
Consuming Miami or New Orleans Drinking Water	7
Lightning	3
Hurricanes	3
Asbestos in School Buildings	1

Source: Harvard report, "Summary of Symposium on Health Aspects of Exposure to Asbestos in Buildings." Based on Sources of Risk Estimates: Commins (1985), Weill and Hughes (1986), Wilson and Crouch (1982)

viruses. Bioaerosols are also discussed in literature addressing microbiological, or microbial, contaminants.

Microorganisms associated with normal communicable childhood diseases, such as mumps and measles, or the usual afflictions like flu and "colds" are not nominally treated in IAQ literature.

Sources

Biological growth sources and sites include wet insulation, carpet, ceiling tile, wall coverings, furniture and stagnant water in air conditioners, dehumidifiers, humidifiers, cooling towers, drip pans and cooling coils of air handling units. People, pets, plants, insects may also carry biological agents into a facility or serve as potential sources.

Biological agents may enter the building through outside air intakes. Due to their small size, they may not be filtered out of the airstream with low MERV level efficiency filters (see Chapter 8). Frequently, they settle in the ventilation system itself. If allowed to become wet and dirty, air conditioning ductwork can provide a fertile territory for bioaerosols. The dust and darkness inside duct work, plus condensate moisture, all work together to turn the vast surface area into breeding grounds for mold. Porous acoustical lining can worsen the problem when installed too near a moisture source, such as a wet cooling coil. In the case of mold, the role of water in any form—bulk water from leaks, high relative humidity, water vapor, or even microscopic condensate—cannot be emphasized enough. Mold spores are ubiquitous because they are in the outdoor air. Most building component surfaces have sufficient nutrient content to enable mold to reproduce. Thus, mold can amplify within a few hours on virtually any interior substrate surface in the presence of sufficient water.

Because of the emerging emphasis and, in some cases, hysteria regarding mold, the subject is covered in greater depth in this edition in each of those sections in the book dealing with contaminant causes, symptoms, and control. It is not dealt with in a separate section or stand-alone chapter since it is but one of a number of contaminants that must be investigated, causes corrected, and remediation performed and many of the elements are similar to other contaminants of concern. Further, Sick Building Syndrome is seldom brought on by a single issue or contaminant of concern. Thus, the professional must perform the evaluation, problem classification, and contaminant control remediation in an integrated and holistic manner.

Symptoms and Health Effects

Common symptoms associated with biological contaminants include allergic reaction sneezing, watery eyes, coughing, shortness of breath, dizziness, lethargy, fever and digestive problems.

Different causes prompt diverse symptoms and conditions. Concerns may range from odors and "stuffiness" to Legionnaire's disease. Little evidence presently exists to support the contention that bioaerosols are responsible for medical problems related to reproductive systems.

Building related illnesses (BRI) constitute a range of hypersensitivity and infectious diseases. Hypersensitivity diseases, such as asthma, humidifier fever and hypersensitivity pneumonitis, are caused by immunological sensitization to bioaerosols. Prolonged exposure to mold spore allergens, for example, wears down the immune system and increases the individual's sensitivity prompting reactions to lower concentrations. One case of hypersensitivity pneumonitis is sufficient to suggest the sensitization process may be occurring in others, and mitigation procedures should be taken. In any suspected building-related illness, a physician should be involved in the diagnosis and the etiology. Unfortunately, an outbreak occurred in Cleveland in the early 90s involving infant fatalities in low income housing. The CDC researchers that investigated the outbreak initially and inappropriately blamed the deaths on mold, specifically *Stachybotrys chartarum*. This species is a black slimy mold that occurs only after prolonged and extensive water damage (that had occurred in the poorly maintained housing in question). The mis-diagnosis was not corrected by CDC until well after the media had created a frenzy about "Toxic Black Mold that can kill kids." The "Toxic Mold" label has stuck even though most medical scientists have since then denied the dire and extreme health affects attributed to "Toxic Mold". Yet, they hold that any of the atmospheric molds in high concentrations or extensive exposure can bring on the general symptoms and allergic response discussed above. At this writing, however, the science is still not clear that mold deserves the negative hype that is being promulgated by the plaintiff lawyers and remediation contractors. Ronald Gots, M.D., writing for ASHRAE IAQ Applications concluded that "While exposure to very high doses can adversely impact a variety of metabolic processes in animals and humans, there is no strong scientific support in the current literature to suggest that adverse health effect... such as brain damage, cancer, chronic fatigue syndrome, or fibromyalgia... are caused by low-dose inhalation exposures potentially encountered in commercial and residential buildings." This is the critical, widespread misunderstanding about mold toxins.

If a health complaint involves an infectious disease; e.g., hypersensitive pneumonitis or Legionnaire's disease, it can be assumed that a bioaerosol is involved.

COMBUSTION PRODUCTS

The major categories of products resulting from combustion can be listed as carbon monoxide (CO), nitrogen oxides (NO_x), particulate material, and polynuclear aromatic hydrocarbons (PAH).

Carbon monoxide (CO) is an odorless, tasteless and colorless gas.

Nitrogen oxides (NO_x) includes nitrogen compounds NO, NO_2, N_2O, OONO, ON(O)O, N_2O_4 and N_2O_5. All are irritant gases, which can impact on human health.

Particulates represents a broad class of chemical and physical particles, including liquid droplets. Combustion conditions can affect the number, particle size and chemical speciation of the particles.

Polynuclear aromatic hydrocarbons (PAH) concentrations are usually low indoors. PAH concerns stem from their potential to act synergistically, antagonistically or in an additive fashion with each other and other contaminants. The chemical composition and concentrations of these compounds vary with combustion conditions.

Sources

Combustion products are released under conditions where incomplete combustion can occur, including: wood, gas and coal stoves, heaters and cooking surfaces; unvented kerosene heaters or appliances; unvented grilles; portable generators; fireplaces under downdraft conditions; and environmental tobacco smoke. Vehicle exhaust is a primary source, particularly from underground or attached garages, as well as from the outdoor air.

Health Effects

The impact on human health varies with the category of combustion product; so they are treated separately below.

Carbon monoxide (CO) has about 250 times the affinity for hemoglobin than oxygen has. When carboxyhemoglobin (COHb) is formed, it reduces the hemoglobin available to carry oxygen to body tissues. CO, therefore, acts as an asphyxiating agent. Common symptoms are dizziness, dull headache, nausea, ringing in the ears and pounding of the heart. Should CO inhalation induce unconsciousness, damage to the central nervous system, the brain and the circulatory system may occur. Acute exposure can be fatal. Young children and persons with asthma, anemia, heart and hypermetabolic diseases are more susceptible.

The extent to which *nitrogen oxides* (NO_x) affect human health is unclear. The most information is available about nitrogen dioxide (NO_2). NO_2 symptoms are irritation to eyes, nose and throat, respiratory infections and some lung impairment. Altered lung function and acute respiratory symptoms and illness have been observed in controlled human exposure studies and in epidemiological studies of homes using gas stoves. Studies in the United States and Britain have found that children exposed to elevated levels of NO_2 have twice the incidence of respiratory illness as children not exposed.

Combustion *particulates* can affect lung function. The smaller respirable particles (less than 2.5 micron μm in size) present a greater risk as they are taken deeper into the lungs. Particles may serve as carriers of contaminants, such as PAH, or as mechanical irritants that interact with chemical contaminants. Respirable particulates, as a contaminant group, are discussed later in the chapter.

The health effects of *polynuclear aromatic hydrocarbons* (PAH) are very difficult to determine or predict. PAH's propensity to act in concert with other contaminants complicates any effort to attribute singular cause and effect. It is known that some PAHs are carcinogens while others exhibit co-carcinogenic potential.

ENVIRONMENTAL TOBACCO SMOKE

Environmental tobacco smoke (ETS) comes from the sidestream smoke emitted from the burning end of cigarettes, cigars and pipes and secondhand smoke exhaled by smokers. Breathing in ambient ETS is generally referred to as passive, or involuntary, smoking.

ETS contains a mixture of irritating gases and carcinogenic tar particles. Because tobacco doesn't burn completely, other contaminants are given off, including sulfur dioxide, ammonia, nitrogen oxides, vinyl chloride, hydrogen cyanide, formaldehyde, radionuclides, benzene and arsenic.

Source

A lighted cigarette gives off approximately 4,700 chemical compounds. The EPA has estimated that 467,000 tons of tobacco are burned indoors each year.

Benzene, a known human carcinogen, is generally emitted from synthetic fibers, plastics and some cleaning solutions; however, the most important exposure is from cigarettes. Benzene levels have been found to be 30 to 50 percent higher in homes with smokers than in nonsmoking households.

Carbon monoxide, nicotine and tar particles have also been identified as the chemicals most apt to impact on health.

Symptoms and Health Effects

In 1985, three federal bodies independently arrived at the same conclusion: passive smoking significantly increases the risk of lung cancer in adults. The Surgeon General warned: "A substantial number of the lung cancer deaths that occur among nonsmokers can be attributed to involuntary smoking." More recently tobacco smoke has been classified as a carcinogen by the International Agency for Research on Cancer (IARC), the Surgeon General, NIOSH, and the USEPA. In its proposed ruling, OSHA provided substantive documentation of the fatalities resulting from ETS. Further, OSHA implicated ETS with many diseases, such as cancer and heart disease and low birth weights, that are referenced in their risk analysis.

According to the EPA, tobacco smoke contains 43 carcinogenic compounds. The agency also notes that ETS is a major source of mutagenic substance; i.e., compounds that cause permanent changes in the genetic material of cells.

Studies have shown that passive smoking significantly increases respiratory illness in children. Asthmatic children are particularly at risk according to EPA and the ALA.

FORMALDEHYDE

Formaldehyde (HCHO) is a volatile organic compound (VOC). Formaldehyde is a ubiquitous chemical used in a wide variety of products and is most frequently introduced into the building during initial construction or renovation. It is a colorless gas at room temperature and has a pungent odor in higher concentrations. Its rate of outgassing can be influenced by increased temperature and humidity.

Source

Formaldehyde is used in many building products, including plywood, paneling, particleboard, fiberboard, fabric, urea-formaldehyde foam insulation, adhesives, fiberglass and wallboard. It acts as a polymeric binder and is also a dirt, stain, and wrinkle retardant. Among potential sources are furniture, shelving, partitions, ceiling tiles, wall coverings, draperies, upholstery, carpet backing and ceiling tiles. Concentrations tend to be highest in prefabricated homes, ranging from 0.1 to 5.0 mg/m^3 (1 mg/m^3 = 0.8 13 ppm HCHO). More commonly the range is 0.1 to 1.0 mg/m.

Symptoms and Health Effects

Clinical and epidemiological data indicate human response to formaldehyde can vary greatly. Some people exhibit hypersensitive reactions. Acute exposure to formaldehyde can result in eye, ear, nose and throat irritation, coughing and wheezing, fatigue, skin rash and severe allergic reactions. It is a highly reactive chemical that combines with protein and can cause allergic contact dermatitis. Table 3-4 shows the effect of short term exposure with ranges and median responses, excluding immunosensitive populations.

Controversy remains regarding the carcinogenic role of formaldehyde in humans. The incidence of cancer in rodents makes formaldehyde highly suspect as a human carcinogen. EPA has conducted research which suggests formaldehyde may cause a rare form of throat cancer in long-term occupants of mobile homes. Chamber studies have shown that a given concentration of formaldehyde may evoke quite different degrees of irritation, depending upon duration of exposure, fluctuations in concentrations and the presence of other agents in the air. As recent as June 2010, EPA issued a draft report declaring formaldehyde a "known" rather than "probable" human carcinogen. The report also cites other adverse effects, including sensory irritation of the eyes, nose, and throat; upper respiratory tract pathology; pulmonary function; asthma and atopy; neurologic and behavioral toxicity; reproductive and developmental toxicity; and immunological toxicity.

**Table 3-4. Effect of Formaldehyde in Humans after
Short-term Respiratory Exposure**

Reported Ranges*	Estimated Median*	Effect
0.06-1.2	0.1	Odor threshold 50% of people
0.01-1.9	0.5	Eye irritation threshold
0.1-3.1	0.6	Throat irritation threshold
2.5-3.7	3.1	Biting sensation in nose, eye
5-6.7	5.6	Tearing eyes, long term lung effects
12-25	17.8	Tolerable for 30 minutes with strong flow of tears lasting 1 hour
37-60	37.5	Inflammation of (pneumonitis), edema, respiratory distress: danger to life
60-125	—	Death

* Concentrations in mg/m^3; I mg/m^3=0.813 ppm
Source: National Center for Toxicological Research (1984)

RADON

Radon is an odorless, colorless gas that is always present at various concentrations in the air. Radon is formed from the decay of radium, which in turn results from the decay of uranium. Radon (Rn-222) and more specifically, the radioactive elements into which it decays (radon daughters) represent a major indoor air concern, particularly in homes.

Radon "daughters" are charged particles and they, in turn, adsorb or attach to solid particles in the air. Approximately 90 percent of the radon daughters attach to larger airborne particles on surfaces before they can be inhaled. The remaining 10 percent represent a significant source of exposure; for, as smaller particles, they deliver a dose to critical lung cells. The following section on respirable particulates discusses this aspect further. About 30 percent of the inhaled daughters are deposited in the lung.

Radon has a radiological half-life of 3.8 days, while the daughters' half-life is about 30 minutes. This rapid decay means the daughters emit high radiation energy levels to a comparatively small volume of tissue; and, in the process, provide a major source of injury to tissues.

Radon exhibits daily and seasonal variation in concentration. Fleming and others have found extreme variability, 7.5 to 25 picoCuries per liter (pCi/L) in a diurnal cycles in many buildings that do not coincide with ventilation rates. Indoor radon concentrations in most climates are much lower in the summer. This is usually attributed to more open windows and doors, which increases air flow and tends to equalize pressure differentials.

Source

Some radon will enter a facility through the water system and off-gassing of building materials; however, the principal source of radon is the soil. Radon typically enters through cracks, voids or other openings in the foundation.

Conditions affecting the flow of radon into a facility are:

•   soil factors—level of radon concentration, emanation rate, diffusion length, permeability, soil moisture;

•   building factors—type and formation of the foundation or substructure, construction quality, design; and

- pressure differentials—building depressurization through stack (thermal) effect and/or exhaust fans, HVAC behavior wind, barometric pressure changes, pressure gradients in the soil.

Symptoms and Health Effects

No immediate symptoms are associated with radon exposure.

When absorbed into the lung cavity, radon decay products may increase the incidence of lung cancer. There is evidence that tobacco smokers exposed to radon are more likely to get lung cancer.

RESPIRABLE PARTICULATES

Particulates represent a broad class of chemical and physical contaminants found in the air as discrete particles. Respirable particulates are generally defined as 10 microns or less in size (PM 10), although EPA has expressed concern that it is the less that 2.5 micron size (PM 2.5) that may be the primary cause of problems in deep lung penetration. This has brought about a revision in the National Ambient Air Quality Standard (NAAQS) to control the average concentrations of PM 2.5. The NAAQS limits particulate matter (PM) in the PM 10 category remains at 150 micrograms per cubic meter for 24 hr. average, and the annual average has been revoked by EPA. The new category of PM 2.5 is limited to 15 micrograms per cubic meter for annual average and 35 micrograms per cubic meter for the 24 hr. average.

Particles fall into two categories; biological and non-biological. Size determines the magnitude of risk and the ultimate residual location in the lungs. Since smaller particles are breathed deeper into the lungs, they can bypass respiratory defense mechanisms thereby creating a greater health hazard. Smaller particles also stay suspended longer and, therefore, offer a greater possibility of inhalation. Smaller particles also offer a greater surface area to total mass ratio. Since respirable particles can also serve as the carriers for other contaminants, such as pesticides, PCBs, radon daughters and pathogens, they can deliver harmful substances to more vulnerable areas. (*Particles associated with specific contaminants, such as asbestos and bioaerosols, are treated in separate sections.*)

Sources

The external environment is a major source of respirable sized particles (RSP) because of ambient pollution. This is because they occur as a result of chemical reaction (i.e. internal combustion engine or forest fire) or phase change (i.e. evaporation of salt water). The introduction of this source is through ventilation, infiltration, and occupant traffic. Biological particles include microbial particles, which may emit harmful organic gases in the air. Plant and animal material also supply biological particles. Common sources of respirable particulates are environmental tobacco smoke, kerosene heaters, humidifiers, wood stoves and fireplaces. Non-biological particles, including dust and dirt, are brought in on occupant clothing, come from occupant activities, maintenance products and the natural deterioration of building products and furnishings. Office equipment such as copiers can also generate RSP from the printer toner.

Symptoms and Health Effects

Irritation and infections in the respiratory track and eye irritation are all symptoms associated with respirable particles. All symptoms and health effects associated with environmental tobacco smoke also apply. Respirable particulates are also associated with lung cancer.

Amman *et al.* listed the following concerns related to respirable particles:

(1)   chemical or mechanical irritation of tissues, including nerve endings at the site of deposition,
(2)   impairment of respiratory mechanics,
(3)   aggravation of existing respiratory or cardiovascular diseases,
(4)   reduction in particle clearance and other host defense mechanisms,
(5)   impact on host immune system,
(6)   morphologic changes of lung tissues, and
(7)   carcinogenesis.

Health consequences vary with the size, mass, concentration and other contaminants acting in concert with the particles. EPA has found that respirable particles at concentrations of 250 to 350 $\mu g/m$ increase respiratory symptoms in compromised individuals.

Because of their adsorption properties, particles carry semivolatile chemicals, such as pesticides, dioxins, and PCBs into humans as they inhale or ingest them. Health effects normally associated with these chemicals, including cancer, can be attributed to respirable particles as well.

Fibrous Glass

Man-made mineral wool and fibrous glass products are widely used in the construction industry. In the HVAC system, fibrous glass is used as insulation and acoustical liner for the distribution system as well as insulation for the air handler and specific components such as terminal boxes and VAV boxes. These components came under some somewhat hysterical scrutiny in the hay-days of AHERA and asbestos removal with the fear that glass fibers would offer potentially the same kind of problem as asbestos. Several factors have defused this initial hysteria. First, we now know much more about fibrous glass than asbestos and asbestos installations. The Washington Post commented in a synoptic article about fibrous glass "Fiberglass is one of the most studied substances on Earth." The actual field data shows that installed fibrous glass systems demonstrate no more glass fibers in the air stream than in background ambient outdoor air. Secondly, this air constituent has been more widely studied by epidemiologists for many more years and with much higher populations than the original asbestos population data. This has provided the manufacturers with much more reliable health related data. More recent data has also shown that the basic man made mineral fiber behaves differently that the natural asbestos fiber. First, its physical size ranges and properties are different. Although, it can be found in respirable size range, it tends to be larger and the fiber tends to break across rather than longitudinally. This avoids the splintering typical of asbestos, which avoids fragments of the fibers from reaching the deep alveoli of the lungs. Further, the fiber is less durable from a chemical standpoint. This means that even if fibers do reach the deep lungs, they don't stay in the body as they are dissolved in the metabolic fluids of the lung tissue. Thus, they do not remain in the tissue to aggravate a wound site and bring about the body's anti-foreign body resistance mechanism. This has led one researcher, Enterline, to conclude that the risk from fibrous glass fibers is "not significant." In an extensive literature search of 350 citations, Woods concluded: "fibrous glass products can be used safely to provide thermal and acoustical benefits if care is provided to maintain these materials in clean and dry condition throughout the construction process and during operation."

Fibrous glass products can cause skin irritation, especially for production workers and installers who have greater physical contact with these products. IAQ professionals should take normal precautions as recommended by the manufacturers when coming into extensive contact with fibrous glass systems. However, the IARC de-listed fiberglass from potential carcinogenicity in early 2003 in an announcement that states "the more commonly used vitreous fiber wools including installation glass wool are now considered not classifiable as to carcinogenicity to humans."

Natural Latex

A rather narrow and specific health effect from particulates is derived from the allergic reaction to natural rubber allergens. Although natural rubber is a ubiquitous component of a wide array of health care and houseware products, the primary focus is on natural latex gloves. This is because of their popularity in health care settings to act as a barrier to exposure to blood-borne pathogens such as HIV. A large number of individuals have exhibited reactions to latex varying from irritation to allergic dermatitis to hypersensitivity. A further concern is that the potent latex allergens are carried in the airstream on fine talc/starch particles used as lubricants in the gloves. This has led to a NIOSH alert that documents this issue and provides recommendations for controlling further exposure of workers in the health care field.

VOLATILE ORGANIC COMPOUNDS (VOCs) AND OTHER GASES

Organic compounds that exist as a gas, or can easily off-gas under normal room temperatures and relative humidity, are considered volatile. A range of VOCs are always present in indoor air.

NOTE: *Formaldehyde is a VOC by definition, but is one of the most widely used chemicals in the construction process. Thus, it is discussed separately. Tobacco smoke and other combustion contaminants emit VOCs such as benzene, phenols. For more specifics, please see other sections in this chapter.*

Sources

Hundreds, if not thousands, of VOCs are found in the indoor air. The list of potential sources is lengthy and growing. Some of the major, and more common sources, are photocopying materials, paints and varnishes, gasoline, people, refrigerants, personal hygiene and cosmetic products, building materials, biological matter, molded plastic containers, disinfectants, cleaning products, environmental tobacco smoke, and even office supplies. Some of these sources emit several VOCs. For example, tobacco smoke contains such VOCs as alcohols, acetone, benzene, formaldehyde, phenols, ammonia, aromatic hydrocarbons and toluene. Few are unique to any single source, since toluene, for instance, can also be found in gasoline, paints, adhesives and solvents.

Symptoms and Health Effects

Symptoms attributable to VOCs include respiratory distress, sore throat, eye irritation, nausea, drowsiness, fatigue, headaches and general malaise.

Specific VOCs are not often proven to cause SBS complaints. Due to the large numbers of chemicals found indoors, it is very difficult to establish any causal relationship between health and certain VOCs. Industrial exposure studies have documented respiratory ailments, heart disease, allergic reactions, mutagenicity and cancer to some VOCs. Combinations of certain VOCs are suspected of having synergistic effects and this potential is currently being researched.

Multiple Chemical Sensitivity

"MCS" is one of the more controversial health effect outcomes of VOC exposure that has been reported. It is a complex mixture of both medical and psychological symptoms. It is thought to be brought about by either or both, a short exposure to extremely high concentrations of volatile chemicals, or a long exposure to relatively low levels of similar chemicals. The immediate reaction reportedly comes about because of a spill or significant chemical exposure. Unfortunately, the prolonged effect is an abnormal (and some experts claim irrational) reaction to much lower level exposures to either the same or related chemicals or even other routinely present chemical compounds. The reactions are skin sensitivity, upper respiratory and even systemic and neural complications. These complex and sometimes bizarre reactions have led to substantial controversy in the medical community. They refer to the complaint as "Time Dependent Sensitization" (TDS) and more recently re-labeled as "Idiopathic Environmental Intolerance." A segment of the medical profession denies the physical condition, claiming that the issue is solely psychological. These critics hold that MCS is psychological in origin because no mechanism for cross-sensitization to unrelated chemicals has been found in the immune system. Additionally, the fact that the symptoms can be inconsistent and unpredictable from the same kind of exposure at different times leads to the interpretation that these inconsistencies are evidence of hypochondria. Other medical clinicians affirm the existence of the condition, but admit that the reaction is so severe that fear, trauma, and pain bring on irrational behavior caused by fears of chemical exposure.

Dr. Claudia Miller, a board certified allergist with the University of Texas at San Antonio, is widely published on her work with MCS. Her work has indicated a hypothesis of involvement with the olfactory-limbic areas of the brain where MCS may be caused when nerves mis-fire. This area of the brain appears to be sensitizable and has lead this researcher to conclude a new theory of disease--which she terms "TILT" or Toxicant-induced Loss of Tolerance. Her hypothesis is that TILT may underlie or overlap with sick building syndrome, chronic fatigue syndrome, autism, depression, bipolar disorder, asthma, and host of other chronic conditions.

The fact remains that a growing number of victims of MCS claim that they have been severely impaired by the exposure to chemicals in the indoor environment.

Office Eye Syndrome

Researchers in Scandinavia (Franck and Skov) have identified another potential outcome from VOC exposure in the workplace in the form of eye irritation that they have dubbed "Office Eye Syndrome." The symptoms are redness, itchy and watery eyes, throbbing, excessive dryness, and various forms of headache. The problem is also worse for contact wearers. Their research has identified the cause to be that low level VOC exposure will dissipate or break-up the tear film moisture barrier that protects the eyes from environmental pollutants. Other environmental problems such as glare, excessive particulates, and low humidity can worsen the reaction.

OTHER GASEOUS CONTAMINANTS
Ozone

This unstable gaseous trivalent form of oxygen (O_3) is a very active oxidant and is usually the product of an energy field such as atmospheric lightening or an electrical corona. It is an airborne contaminant listed in the NAAQS standard and is a component of ambient photochemical smog. It can also source from malfunctioning electrical apparatus such as electronic air cleaners. At elevated concentrations, ozone has been employed in industrial or commercial settings to control through oxidation the growth of bacteria, malodors from fires, and reactive volatile organic compounds. Ozone has also demonstrated successful application for contaminant control in liquid phase treatment, such as cooling towers. However, it can be equally destructive of other oxidizable materiel, such as building materials, finishes, as well as the delicate membranes of the upper respiratory region. According to the American Lung Association (ALA), ozone is a potent lung irritant and a concern for human exposure. For this reason, it is considered by EPA to be a contaminant of concern for human exposure at short term exposure concentrations in excess of .08 ppm (or 80 ppb). According to an ALA bulletin on residential air cleaners, ozone generators, ion generators, and other devices yielding excess ozone should be avoided, especially by children, the elderly, and persons with asthma and other lung diseases.

Another aspect of ozone triggers from the overall instability and reactivity of the trivalent oxygen molecule. Being so unstable and reactive, it can quickly decay through cross-reaction with other reactive compounds in the airstream, as found in the collective stew of volatile organic vapors present in the indoor environment. When reacting with prevalent compounds in the aldehyde or amine families, very odorous and/or irritating products of reaction can result. The work of Weschler, et al, has revealed that even low levels of ozone concentrations can result in both unwanted chemical constituents as well as high levels of ultra-fine particulate matter consisting of condensation nuclei. These products of reaction can cause serious defects and failure of delicate electronic equipment as well as pose exposure and potential health risks to occupants.

Reactive Gases

Earlier this decade, a "perfect" storm arose starting with a literal storm—Katrina, which created a tremendous demand for drywall for remediation, restoration, and renovation and on top of an already overheated residential construction market underway. This created shortages and cost increases in drywall that was quickly rectified with imports from the world's favorite manufacturing source, China. Thus, the Chinese Drywall (CDW) drama was underway. It did not take long to find that the CDW came with a disguised and unwelcome bonus. The product was laden with latent corrosive, reactive chemicals that out-gassed when the moisture loving gypsum reached equilibrium. Predominantly sulfur bearing gaseous compounds, such as hydrogen sulfide, converted to acids in contact with moisture which in turn, can react with the wiring, plumbing, ducting, fasteners and other metal components of the built environment. Reportedly, the H_2S outgassing rate in the CDW exceeds US made products by 100 fold.

Much of this attack occurs in concealed space within the walls and structure. Thus, the negative impact of the reactive outgassing from CDW can have devastating results if not found promptly and remediated thoroughly. As of this writing, litigation blog sites report that over 3000 homes are involved in some stage of litigation with a few settlements in the $100-165k range. Reported, sufficient board has been imported to impact up to 50,000 structures. Although consensus mitigation protocols do not exist yet, the immediate prima facie remediation is the removal of all deficit wallboard and the replacement of affected metal systems such as plumbing and wiring. Further experience and research will be required to determine the role of secondary sinks and exposed porous surfaces.

A number of class action suits are underway which may expand these numbers considerably. In response to the need to determine the presence and extent of CDW in the structure, IAQ professionals are employing a variety of hand held diagnostic instrumentation to sense the reactive chemicals. The reader is cautioned that the H_2S is present in very low concentrations in the PPB range so care should be taken to employ instrumentation that employs appropriate sensors with adequate sensitivity for these low concentrations.

Odors

Odors are a class of contaminants in gaseous form that can bring about discomfort, irritation, stress, complaints, and even fear, panic, and mass hysteria. Odor complaints are usually very difficult to track down as they seem to be nebulous, erratic, and very "Will o' the Wisp." One factor contributing to this problem is that airborne chemicals can have dramatically different odor thresholds. Distinctive malodorants like ammonia (smelling salts), Sulfur dioxide (rotten eggs), and Hydrogen sulfide (sewer gas) have dramatic different approximate odor thresholds—NH_3=47ppm; SO_2=.47ppm; and H_2S=.00047ppm. Thus, perception ranges over 5 orders of magnitude. Blends of malodors can equally affect the odor threshold perception.

Seldom are occupants made really ill from odors, but their reaction is often "dis-ease." This results in passionate and vocal complaints. Because of these factors, facility managers often relegate odor complaints to the bottom of the priority pile. This brings even more passionate and seemingly irrational complaint response from occupants. To understand this reaction, it may be helpful to know more about the basics of human odor perception.

Odors are gaseous chemicals that are sensed by chemical receptors located in the upper nasal area and connected to the olfactory nerve. This nerve carries the nasal olfaction message to the brain where it is received in the limbic region. This is the oldest portion of the brain from an evolutionary basis and, thus, it is the area that contains the more basic animal behavior patterns and responses. These include loves, lusts, hates, fears, rages, and remembrances. It is the brain region of emotion and odor is the driver of recognition and response to: home, mother, safe, food, mate, and sex, but also, enemy, fear, threat, unsafe, unpleasant, lost, and bad food. It is also the region of the hypothalamus and the pituitary gland. These control the "fight or flight" reaction and place the body under the positive stress of "battle alert" through adrenaline.

Chemical odors, particularly offensive malodors, are received by the body as negative threats to well being. "Something smells bad in my work space—take care of it now!!!" is a message of unclean, unsafe, unhealthy or bodily threat! This reaction is a "fight or flight" reaction that brings about physical "dis-stress." This is a basic core message that the body is in jeopardy, and explains why odor complaints are passionate, imperative, and even irrational in nature. This is also why odor complaints should be handled promptly and thoroughly, even though their diagnostics is often frustrating and unrewarding. However, another reason that odors are to be handled with high priority is that they may be a "telltale" that provides an instructive clue as to the nature of a brewing IAQ problem. The noxious smell of sewerage may foretell the infiltration of poisonous sewer gas. The annoyance of auto exhaust odors and/ or headaches can provide early warning of carbon monoxide poisoning (which is non-odorous). The odor of "cellar" or "locker room" can foretell a serious moisture and unresolved microbial growth problem.

ENVIRONMENTAL CONDITIONS

Contaminants may not act alone in affecting occupant health. General environmental conditions, such as temperature and lighting, can interact with contaminants. Environmental conditions may act independently, physically, antagonistically or synergistically with various contaminants. For example, room temperature and relative humidity

(RH) have a significant impact on the rate formaldehyde off-gases. RH also dictates the breeding opportunities for various agents. Air that is too humid, for example, may foster the growth of mold.

Factors related to thermal environment conditions are discussed in detail in Chapter 6. Other factors that may serve as a source of IAQ symptoms or interact with the contaminants include artificial light, noise and vibrations, particles and fibers, ions, psychological and ergonomic considerations. These related factors have promulgated the term "IEQ" meaning Indoor Environmental Quality as a more appropriate label for the indoor conditioned environment.

ARTIFICIAL LIGHT

Poor lighting conditions can cause eye strain and irritation as well as headaches and may increase sensitivity to certain contaminants. Visual stress may come from insufficient contrast in the material, brightness, glare and inappropriate light levels.

Brightness

Brightness is determined by the relative amount of light available at the work surface in relation to the level of illumination in the field of view. The eyes functions most comfortably and efficiently when the brightness relationships are not excessive. The light at the desk surface, for example, should not be more than three times the level of light immediately around the desk. A desk lamp which is being used in a dark room exceeds the recommended brightness ratio and reduces eye comfort and efficiency.

Glare

In maintaining a lighting system, conditions creating glare should be avoided. The position of the light source in relation to both the viewed surface and the eye is critical. For example, light in front of a desk that strikes the surface and is reflected into the eye can create considerable eye strain. This effect, called veiling reflection, is shown in Figure 3-2.

A simple way to check for veiling reflection is to place a mirror on the work surface in front of the worker. If the reflection from the mirror strikes the worker's face, then that person is being subjected to unnec-

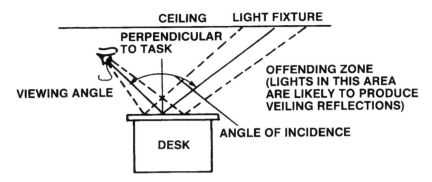

Figure 3-2. Veiling Reflection

essary glare. Light sources that cause this effect should be removed or repositioned; or, in the case of movable work surfaces, the work area should be repositioned in relation to the light.

Quantity of Light

A number of sources discuss lighting and recommended lighting levels. The Electrification Council's manual, *Fundamentals of Commercial and Industrial Lighting*, available through participating utilities, is an excellent source on lighting. The Illuminating Engineering Society of North America (IES) has published recommended levels of lighting in its *IES Handbook, Reference Volume.* The lighting section in ASHRAE 90.1-2007, coauthored by IES, is a good reference. Lamp manufacturers also have information available. For example, Table 3-5 shows the generally recommended levels of illumination in footcandles appropriate for educational facilities.

Some fluorescent lamps have had a negative affect on hyperactive children. Earlier high pressure sodium (HPS) lamps were thought to cause headaches and malaise. Wilkins *et al.* reduced the fluctuation in illumination using a solid state high frequency ballast and decreased eyestrain and headaches by 50 percent in a group of office workers. Recent studies suggest full-spectrum lighting reduces absenteeism. VDT users require particularly well-designed lighting.

NOISE AND VIBRATION

Noise has been found to affect human health by volume, sound pressure levels, infrasound and frequency. Noise at 70-80 decibels (dB)

Table 3-5. Recommended Illumination Levels for Educational Facilities

AREA / ACTIVITY	FOOTCANDLES ON TASK USAGE CATEGORY*		
	I	II	III
Educational facilities			
Classrooms (general)	50	75	100
Science laboratories	50	75	100
Lecture rooms			
Audience	50	75	100
Demonstration	100	150	200
Auditonums			
Assembly	10	15	20
Social activity	5	7.5	10
Music rooms	50	75	100
Sight savings rooms	100	150	200
Study halls	50	75	100
Typing	20	30	50
Corridors	10	15	20
Toilets and washrooms	10	15	20
Food service facilities			
Dining areas			
Cashier	20	30	50
Cleaning	10	15	20
Dining	5		10
Kitchen	50	75	100
Gyms30			
Locker rooms	10	15	20
Offices			
Lobbies, lounges & reception	10	15	20
Offset printing	20	30	50
& duplication area			
General and private	20	30	50
Shops			
Woodworking	20	30	50
Rough bench work	20	30	50
Fine bench work	200	300	500
Maintenance facilities			
Garages-repairs	50	75	100

*Usage categories I, II, and III are determined by weighting factors which consider occupants ages, room surfaces reflectances, speed and/or accuracy of task involved, and reflectance of task background. See *IES Handbook* for complete discussion of method for determining illumination levels.

is suspected of causing tiredness. The Occupational Safety and Health Administration sets industrial guidelines at 90 dB. In comparing two situations where the sound levels were both approximately 60 dB and room features were similar, frequency analysis revealed sound pressure levels in the 8-125 Hz range were much higher in the area described as "unpleasant" by the workers. In situations where levels are above 120 dB, infrasound (sound waves in 0.1-20 Hz range) may cause dizziness and nausea. The more common situation to cause problems is found where industrial machines or ventilation equipment emit low frequency noise (20-100 Hz).

Several smaller studies have found a relationship between health and vibrations. The causative link is believed to be the characteristic resonance frequencies (1-20 Hz) of certain body parts, especially the eye, and external vibrations. A significant correlation was found between office worker irritability and dizziness and the level of vibrations (from a nearby pump room) measured at their desks

The International Standard Organization (ISO) has done extensive work on acoustics and on the effects of vibrations on the human body. Information on their reports, *Exhalation of Human Exposure to Whole-Body Vibration* (1985) and *Acoustics, Description and Measurement of Environmental Noise* (1987), can be obtained by contacting ISO in Geneva. Under the auspices of the Americans with Disabilities Act of 1990 (ADA), the Acoustic Society of America (ASA) has developed noise standards for the school class room with target levels of 35 dBA with reverberation time of .6 seconds republished as ANSI/ASA S12.60-2010.

IONS

With the advent of office equipment requiring high power charges, such as computers, considerable concern has been evidenced about the relationship of negative ions in the atmosphere and SBS. To date, the results are inconclusive.

Guilleman describes negative ionizers as releasing significant amounts of ozone, which is considered a potent throat irritant. In a double blind study, Finnegan *et al.* found the level of symptoms of SBS was not influenced by the concentration of ions in the air.

PSYCHOLOGICAL AND ERGONOMIC FACTORS

A direct causal relationship between psychological factors and SBS has not been established. The World Health Organization reported in 1986 that there is some evidence that stress, which may be the result of some psychological factors, can make individuals more susceptible to environmental factors. Work by Hedge *et al.* and Morris *et al.* while far from conclusive, suggests that SBS may be responsible for stress, rather than the reverse which many have assumed.

There is an increasing body of information regarding the health effects of computer work, particularly monitor screens. There is a consensus that prolonged screen exposure can cause eye irritations, headaches, tiredness and appears to be a factor related to more general IAQ complaints. There has been some indication that the incidence of miscarriage is related to work with screens; however, more research in this area appears warranted.

One of the most telling psychological factors related to the modern work environment may be embodied in a factor listed by McDonald in 1984. He found the fact that workers have little or no control over regulation of temperature, humidity and lighting at their work location was a common problem relative to working conditions.

The effects of ergonomics have been studied in Europe for a number of years. It has more recently emerged as a concern in the US. Defined as the study of problems people have in adjusting to their environment or the science of adapting working conditions to the worker, it generally refers to the physical conditions. Stettman *et al.*, reported that they found ergonomic factors influence the perception of indoor air quality.

HEALTH, CONTAMINANTS
AND ENVIRONMENTAL FACTORS

Table 3-6 summarizes the major SBS factors related to the paradigms offered above to present the relationships of health effects, contaminants and environmental factors. Table 3-7 summarizes the health effects, contaminants and sources associated with building related illnesses.

The data assembled in Tables 3-6 and 3-7 were digested from a careful review of the literature related to SBS and BRI health effects. The tables are by no means inclusive or comprehensive; they only highlight

conditions related to each category. They are certainly not medically definitive. To be as accurate as possible, the terms used by the epidemiologists and others reporting their research were used verbatim; e.g., "dry
cough" and "coughing," and no medical interpretations were attempted.

This material is not intended for medical diagnosis. Rather, the
tables are designed to help the practitioner sort through what is known,
the symptoms, and to discern possible contaminants/sources in Table
3-6 and possible BRIs/contaminants/sources in Table 3-7. An admin-
istrator using Table 3-7, for example, may discern that an occupant appears to have the symptoms in Complex A, which would suggest medical involvement is warranted, bioaerosols are suspect and investigation
of the related primary sources should be considered.

If Tables 3-6 and 3-7 suggest the process is easy or straightforward,
then they are misleading. It is seldom a straight shot from symptom to
contaminant to source. The search, unfortunately, is usually a sorting
process of exclusion. It is confused by the fact that several contaminants
can manifest the same symptoms. Nitrous oxides and formaldehyde, for
example, evidence strikingly similar symptoms. Some dusts and various
microbial contaminants prompt similar allergic type reactions.

Despite the confusion, it is easier working from the known; i.e.,
symptoms, than identify contaminants and trying to find our way back.
In many instances, the causes are never identified and the contaminant-
to-remedy process never even gets started.

A review of the tables will reveal that they merely condense major
findings offered throughout this chapter; so the practitioner can, at a
glance, sense cause/effect relationships that may be at work in a given
facility. To pursue particular contaminants/sources/environmental conditions as they relate to certain health effects, the reader is urged to refer
to earlier portions of this chapter.

In using the tables, the reader should keep in mind that every
health effect cited may be the result of totally unrelated medical conditions not associated with SBS or BRI or a specific workplace.

Table 3-6. The Practitioner's Guide to Health Effects, Contaminants and Environmental Factors (Continued)

SYMPTOM COMPLEX ILLNESS HEALTH EFFECT	POSSIBLE CONTAMINANTS	PRIMARY SOURCES	ENVIRONMENTAL CONDITIONS
Throat, Lower Respiratory Tract	NO_2	Incomplete combustion — stoves, fireplaces, ETS heaters	Low relative humidity
Dry throat, no inflammation	Formaldehyde	Building products & furnishings	
Shortness of breath without lung infections or bronchial asthma	VOCs	Broad product range (See VOCs section)	
Lung Cancer	ETS	Passive Smoking	
Irritation & infection of respiratory tract	Particulates	Combustion products asbestos	
Headache, Fatigue, Malaise Headaches — frontal afternoon occurrence Poor concentration	Bioaerosols	Ventilation systems, humidifiers, dehumidifiers, wet building products, drip pans,cooling coils in AHU, people, pets, plants, insects, outside air	Ergonomic conditions
Dizziness Tiredness Irritability	VOCs	Broad range (See VOC section)	Noise and vibrations
With nausea, ringing in ears, pounding heart	CO	Incomplete combustion — vehicle exhaust — stoves, fireplaces, ETS — gas appliances, heaters, outside air	
Fatigue	Formaldehyde	Building products & furnishings	
Skin problems dryness, irritation			Warm air, Low relative humidity Excessive air movement
Rashes (improves away from building)	Formaldehyde Fiberglass, bioaerosols	Building products and furnishings, insulation, fungal growth, VOCs	

Table 3-6. The Practitioner's Guide to Health Effects, Contaminants and Environmental Factors

Sick Building Syndrome

SYMPTOM COMPLEX ILLNESS HEALTH EFFECT	POSSIBLE CONTAMINANTS	PRIMARY SOURCES	ENVIRONMENTAL CONDITIONS
Eye Irritation Burning, dry gritty eye without inflammation	NO_2	Incomplete Combustion —stoves, fireplaces, ETS	Artificial light Low relative humidity
	Formaldehyde	Building products & furnishings	
	VOCs	Broad range of products (See VOC section)	
Watery eyes	Bioaerosols	Ventilation systems, humidifiers, dehumidifiers wet insulation, drip pans, cooling coils in AHUs, people pets, plants, insects, outside air	
	Particulates	Combustion products, ETS, dust, dirt, maintenance products, building product deterioration, outside air	
Nasal manifestations "Stuffiness" Nasal irritations, rhinorrhoea	NO_2	Incomplete combustion — stoves, fireplaces, ETS heaters	Low relative humidity High temperatures
	Formaldehyde	Building products & furnishings	
	Bioaerosols	Ventilation systems, humidifiers, dehumidifiers; wet building products, drip pans, cooling coils in AHUs, people, pets, plants, insects, outside air	

Table 3-7. The Practitioner's Guide to Health Effects, Contaminants and Environmental Factors (Continued)

SYMPTOMS	POSSIBLE BRIs	POSSIBLE CONTAMINANTS	PRIMARY SOURCES
COMPLEX C— headache* - fever & chills* - dry cough* - chest pains, shortness of breath -vomiting, abdominal pain - diarrhea - weight loss - myalgia - pneumonia	Legionnaire's Disease (LD)	Legionella pneumophila	Aerosols from cooling towers, evaporative condensers, shower heads, water faucets, hot water systems, Jacuzzi, water fountain, waterfalls, hydro-therapy units, stagnant water
COMPLEX D Milder form of LD No pneumonia	Pontiac fever	Legionella	Aerosols from cooling towers, evaporative condensers, showers, water systems, stagnant water
COMPLEX E (Infectious) - headaches* - fever & chills* - cough* - respiratory infections - broad range of other symptoms	Fungal infections Viral infections	Over 100,000 fungi have been described. Of interest are aspergillus, Stachybotys, and mycotoxins. Molds and spores, or saprophytic fungi (about 300,000 species) Viruses	Outside air, duct work People, pets, insects, plants Water damaged materials, organic sources People, pets
HEALTH EFFECTS LONG TERM No immediate symptoms	Asbestosis Mesothelioma & lung cancer Lung cancer	High level exposure to asbestos fibers Primarily amphiboles asbestos, not common in U.S. Radon, radon daughters Particulates	Primary industrial exposure Asbestos Containing Materials (ACM) Soil (water, building materials) Combustion product

*Symptoms common to all BRIs

Table 3-7. The Practitioner's Guide to Health Effects, Contaminants and Environmental Factors

Building-related Illnesses

SYMPTOMS	POSSIBLE BRIs	POSSIBLE CONTAMINANTS	PRIMARY SOURCES
Rhinitis Asthma—like symptoms Other allergic reactions	Allergies	Allergens — microbial	Poor maintenance Humidifiers, esp. cold spray
		Allergens — chemical e.g., formaldehyde	Building products & furnishings
SYMPTOM COMPLEXES			
COMPLEX A - headaches* - fever & chills* - dry cough * - lung disorders - malaise & myalgia - shortness of breath (worsens during exposure)	Hypersensitivity pnuemonitis (extrinsic allergic alveolitis)	Organic dust (microorganisms, endospores, animal protein) Low molecular weight chemicals	People, pets, plants, insects Outside air Ventilation systems, humidffiers, dehumidifiers, wet insulation, drip pans, cooling coils m AHUs
COMPLEX B - fever & chills* - coughing* - lethargy - arthralgic & myalgia - polyuria - weight loss - breathlessness (lessens during exposure)	Humidifier fever	Microorganisms (imcriminated organisms not yet indisputably identified)	Humidifier

Chapter 4

INVESTIGATING INDOOR AIR PROBLEMS: HOW TO FIND OUT WHAT WENT WRONG

When indoor air problems mushroom into full scale concerns, owners and operators have a tendency to turn immediately to outside expertise. It is appropriate to respond immediately; however, a lot can be done through simple in-house steps before such a move becomes necessary. In fact, a walk through survey and a few corrective actions may avoid the call for outside consultation entirely. Should it become necessary to engage an IAQ diagnostic team, the preliminary work by the staff can greatly expedite the team's efforts.

When a building is "sick," two steps can be taken before the "doctor" arrives. Staying with the medical analogy, the first step is to "Take two aspirin and call me in the morning;" i.e., do the simple things in the hope that the doctor will not be needed in the morning. The second step is "What to do until the doctor comes."

If the "doctor" is needed, treatment may be a simple prescription, or the doctor may find that some extensive tests are required. In either case, some cataloging of symptoms, their patterns and building conditions can prove helpful.

If symptoms appear to be serious, life-threatening, or are likely to cause long-term health damages, a diagnostic team should be brought in immediately! The investigations should proceed as rapidly as possible. Immediate steps should be taken to protect the occupants in such a situation, including evacuation of the area or the building as appropriate.

It should be stressed that the investigation process does not replace a disaster plan, such as procedures to respond to a chemical spill in or near a facility. Nor does it address long term hazards.

From the simple inspection to complex diagnostics and post-treatment monitoring, IAQ building investigations can be divided into the following stages:

Phase I   *Preliminary Assessment*—a self-evaluation, data gathering, observation effort.

Phase II   *The Quantitative Walk Through Inspection*—conducted by trained in-house staff, or as a preliminary inspection by the diagnostic team. Measurements are generally confined to the use of a smoke pencil and single gauges for temperature and humidity. Many IAQ problems (up to 80% according to some investigators) are usually identified by this stage of the investigation process.

Phase III   *Simple Quantitative Diagnostics*—more extensive analytical procedures conducted by the diagnostic team; limited measurements of implicated factors or surrogates. Some investigators perform these procedures during the Phase II walk through.

Phase IV   *Complex Quantitative Diagnostics*—broad in-depth testing; qualitative studies of factors in combination; medical examinations.

Phase V   *Proactive Monitoring and Recurrence Prevention*—observation, record keeping, re-testing as warranted; preventive measures.

With a little background information and training, the first phase can easily be handled internally. Depending on the staff's level of expertise, in-house personnel may also conduct part, or all, of the walk through inspection. Outside consulting professionals may be useful for initial training of in-house staff. Even with trained staff, however, it may be advisable to get outside consultation should the emotional climate or the seriousness of the IAQ problem warrant it. Complex Quantitative Diagnostics require the special expertise of a multidisciplinary or multi-experienced team.

Management concerns related to problem investigation focus on in-house investigation strategies associated with Phases I and II. If the diagnostic phases become necessary, management responsibilities then extend to data preparation and consultant selection, investigation support and oversight, and communication with affected parties. In order to appreciate the work that will be required of a diagnostic team, management needs a general understanding of diagnostic procedures. The more detailed technical and contaminant-specific procedures do not generally fall within the management purview and are not treated in this text.

Each phase of an investigation is apt to suggest remedies. Before proceeding to the next phase, corrective measures should be taken and the effects of those measures observed. The move to the succeeding phase(s) will depend on how well the preceding efforts have worked in alleviating the problem(s). The progression of the investigative phases, appropriate personnel related actions and evaluation procedures are shown in Figure 4-1. Note that each phase is followed by monitoring procedures and a decision as to whether or not the investigation needs to progress to the next phase.

INVESTIGATION PROCEDURES

As with any new field, there seems to be an abundance of people, who are prone to embroider and embellish IAQ concerns with Latin phrases and technical jargon. Others get caught up in their fields of expertise.

Medical confirmation supplied a couple months later may be interesting, but hardly vital to the building manager who has the problem *now.*

It is not necessary to know such technical information as "Q fever is a zoonosis caused by *Coxiella burnetti,*" to get to work. Scientists want to know specific causes; managers want solutions.

Where to Start

The best investigation procedure currently available is the solution-oriented approach. Since there is so much we still don't know about detecting indoor air problems and so many difficulties associated with measurement and verification procedures, the simplest process is to start with what we do know and chip away at it. The important thing is to start the response process immediately.

With the knowledge afforded from other sections of this book, it may be possible to spot the problem almost immediately and fix it. If

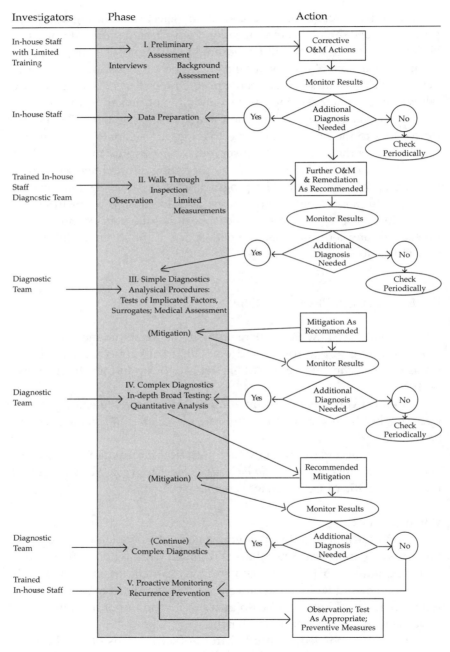

Figure 4-1. Investigative Phases

that is the case, a place on the complaint form (refer to the Management Procedures Chapter) can be used to indicate what was done, when it was done, who did it and any follow-up monitoring that is needed.

If, however, the cause is not immediately obvious, or a broader investigation seems appropriate, a preliminary assessment of the situation should be the first step.

A prompt site visit not only aids in identifying many sources, it can reduce the spread of the illness, contain economic losses and demonstrate a caring attitude toward the affected people. Delays can potentially increase exposure time and worsen the health effects while exacerbating the core causes of the problem. Delays aggravate inspection difficulties as physical changes are frequently compounded over time. Irritated personnel, who feel their concerns have been ignored, can cloud or exaggerate the situation. Early recognition of a problem as well as timely and systematic evaluation are key factors to a quick and effective resolution to both the physical, as well as the psychological problems.

Phase I: Preliminary Assessment

The Preliminary Assessment is designed as a first response to reports of indoor air problems. It is a data gathering and observation effort. No measurements are taken. No professional consultation is required. It is important to resist the temptation to respond to the "Come test my air—something is wrong!" comment. Testing without the guidance of the preliminary assessment and elements of the walk-through can be a waste of money and effort. It can yield information without knowledge or understanding that can introduce false negatives or false positives. Furthermore it becomes a "red herring" and that delays the process of gaining real insight into causes and solutions.

The purpose of the Preliminary Assessment is two fold: (1) to get a better understanding of the extent of the problem and related conditions; and (2) to identify possible causes. The assessment procedure has two distinct facets, determining: (a) the nature and scope of the complaints; and (b) a preliminary audit of facility conditions, systems, maintenance and operational procedures.

The observations are best done by a team. Someone on the team should have knowledge of HVAC system operations. It is valuable to have someone from outside the facility (not necessarily outside the organization), who can take a fresh look at conditions. An effective preliminary assessment process does require that those conducting the assessment have some knowledge of indoor air quality concerns and acceptable data gathering procedures. The training need not be extensive, but effective self-evaluation needs some guidance.

The areas examined by the team should be influenced by the nature, timing and locality of the complaints. If an occupant has been medically diagnosed as having hypersensitivity pneumonitis, the contaminant is a bioaerosol and testing for formaldehyde would not be the first order of business. This is not to say that a broader investigation may not be warranted if problems persist, but the order of the day is to find the source of the identified problem.

Malfunctions in the ventilation system may have allowed concentrations of a contaminant to reach the level that affects health. During the Preliminary Assessment, a *visual* check of the system should be made to be sure it is performing satisfactorily.

Caution is warranted: The Preliminary Assessment phase is not intended to, nor will it, resolve many indoor air problems. It is as its name describes, a *preliminary* effort. Neither are the procedures designed to treat indoor air problems that are not manifested in SBS symptoms, such as asbestos and radon. These contaminants must be treated by inspection and control procedures discussed in Chapter 7.

Nature and Scope of the Complaints

Determining the extent of the problem will usually involve some preliminary interviews to discover the area(s) in which the symptoms are evidenced, the pattern, the duration and the number of people experiencing similar symptoms or medical signs. Psychological and ergonomic factors may also need to be considered. Some of this information may be available through a review of IAQ complaint forms. Care should be taken in the interviewing process to not inflame the situation further, as widespread interviews or written questionnaires can result in undue attention to the issue and can agitate or aggravate more issues than they resolve. Wherever possible, therefore, the needed information should be gathered from existing human resources records, health records, or complaint logs. This avoids further aggravation of productivity and human relations issues. Symptom and background information an investigator should seek would include:

How long have you worked in the area?

When did you first notice the symptom?

What symptoms have you experienced? (headaches, chest pains, shortness of breath, nose and/or throat irritation, nausea, flu-like symptoms, abdominal pains, drowsiness, lethargy, fatigue, fever, eye irritation, skin rash or itching, etc.)

Do you suffer from any current or prior medical problems that may be related to these symptoms? (asthma, allergies, hay fever, eczema, migraines, arthritis)

Do you take any medication? (over the counter or prescribed)

How often and when do the symptoms occur? (times per day, week, month)

Do the symptoms occur during a particular part of the year? (winter) Week? (Monday morning) Or time of day? (First thing in the morning, end of the day)

How long do they last? (all morning, all day, all week, all the time)

Are there any particular areas in which you work where the problem seems greater?

Are there any specific tasks you perform just before symptoms are noted?

Do symptoms go away upon leaving the building?

Do you wear contact lenses? Operate visual display terminals or copiers more than 10 percent of the day? Operate any special equipment?

Do you smoke? Are you bothered by others smoking in designated areas?

How would you describe the environment where you work? (noticeable odors, too hot, too cold, stuffy, dusty, noisy, etc.)

An interview form, which addresses these matters, is presented in the Appendices although great care should be taken by managers in the use of this process as to the method that questions are asked and interpreted.

Review of the data regarding the symptoms may reveal patterns of coincidence as to symptom cluster, timing, seasonality, or location. To the extent that the symptoms suggest a probable source, the facility inspection may then focus on that source. For example, in classifying IAQ problems, a very distinctive "Monday Morning Phenomena" was discussed, which is associated with humidifier fever. If the interviews reveal this symptomatology, then at the top of the facility inspection list should be humidifiers.

Interviews and Questionnaires

Talking to the people who have registered complaints *and those who have not* can be a valuable means of securing information about the nature and scope of the problem. Non-complainants can help pinpoint odor or temperature problems. They can also help establish that some symptoms may not be building generated. Care should be taken in the interview process to obtain objective data and avoid biasing the data by suggestion.

The usage of widespread survey questionnaires should be very carefully considered. Often, this process only fans the fires of misinformation and mistrust and provides meager and potentially misleading information. At present, there is little consensus as to the value of a questionnaire in the preliminary assessment process. To some degree, this can be attributed to the perspectives of the disciplines involved. Some think questionnaires are indispensable and others have no use for them at all. Just a few thoughts from respected IAQ authorities, will offer a glimpse of the dichotomy. One source says,

> Questionnaires have been shown to be powerful tools in screening and clinical diagnosis. They also play an important role in indoor air quality investigations.

The National Institute of Occupational Safety and Health (NIOSH) has found that when the questionnaire is administered prior to the initial visit, the findings enable the NIOSH team to develop more effective strategies in dealing with the problem and to use of the investigators' time on site more efficiently.

On the other hand, others point out that while questionnaires may be valuable tools for epidemiologists to establish gross averages, they are easily misapplied in individual situations.

The written questionnaire is a specialized tool for IAQ investigators that can easily be misused by lay persons in obtaining and interpreting IAQ data. If the investigative process is to benefit from the use of questionnaires, they must be administered and interpreted by those skilled in their use. For this reason only, an interview form has been provided in Appendix D.

When used correctly, the interview procedure should help define the complaints, assist in determining if the problem is localized and whether any special circumstances; e.g., activities, time of day or week, improves or worsens the problem.

Background Assessment

Prior to any extensive background assessment, it is best to ascertain whether operational conditions in the affected area are normal, or were abnormal, just prior to the IAQ episode. If complaints have been generated by an abnormality, such as an equipment failure, rectifying the situation may satisfy the problem. Thus, no further assessment may be needed. However, the cause and effect should be noted and recorded in case the complaints reoccur.

Background assessment procedures seek to identify what is, as compared to what ought to be. The basic purpose is to assess the overall condition of the building, what changes have been made and whether systems are operating properly. Original construction information, such as construction dates, square footage, ventilation system design, materials used, etc., is gathered along with any changes that could affect thermal or contaminant loads. A review of changes and modifications helps to identify the possible introduction of contaminants from building materials and furnishings as well as unmet ventilation needs. Inspection procedures should note any changes in space utilization, such as relocation of copiers and laser printers to poorly ventilated areas of the office. The background assessment procedures can also serve a valuable function in verifying the information collected in the interviews.

As information regarding the original design, construction, subsequent changes and use are assembled, the assessment also seeks to identify probable contaminant sources. Years of IAQ investigation has provided a good indication of where those sources are. During the assessment process, an experienced and "educated eye," which is sensitive to probable sources, can recognize many opportunities to improve indoor air quality. This experience is incorporated into the following discussion.

Early Detection of Possible Sources

NIOSH, after hundreds of IAQ building investigations, remains settled on a solution-oriented approach. The approach is one of exclusion. They progressively eliminate the most likely causes and gradually narrow the range of possible sources. The NIOSH approach is reminiscent of the sculptor, who explained the best way to carve an elephant was to take a piece of stone and chip away everything that doesn't look like an elephant.

A solution-oriented approach can effectively guide the initial steps of the lay person seeking solutions in their own building. Many problems could be averted or resolved in-house if checks were made of the areas where the problems are most likely to exist.

A checklist of the typical sources for contaminants can be compiled from the primary types of problems found by major investigating teams. Even though some IAQ episodes have been found to be multifactorial, for our purposes they can still be classified by the primary type of problem found. A review of the frequent IAQ problems found by early investigators shown in Table 4-1 reveals a commonality that can still serve as a sound basis for early detection of probable causes.

Comparing these lists becomes a process of using what these organizations have learned to your advantage. Ventilation, as a source of indoor air problems, is shown on every list. It makes good sense, therefore, to carefully assess the conditions of the HVAC system, as the building assessment is conducted.

Table 4-1 also reveals that early investigators have found ventilation *distribution* frequently shows up as a problem. The generalized category labels are not as informative as they could be especially with Investigator I, since their "inadequate ventilation" notation really dealt with the entire distribution system. This even included ventilation effectiveness and maintenance issues.

The following items should guide the investigation team through the process of the background assessment. In all likelihood, the same elements will be repeated the in greater detail and depth during the Phase II investigation by the outside consulting team.

Inadequate Ventilation Causes

• Not enough fresh outdoor air supplied to the office space;

• Poor air distribution and mixing which causes stratification, draftiness, and pressure differences between office spaces;

Table 4-1. Sick Building Syndrome Problems

Org.	Investigator 1	Investigator 2	Investigator 3
Bldgs.	529	50	223
Yr	1987	1989	1989
	Inadequate ventilation (52%)	Operations & Maintenance (75%) — energy mgmt. — maintenance	Poor ventilation —no fresh air (35%) — inadequate fresh air (64%)
	Inside contamination (17%)	— changed loads	— distribution (46%)
	Outside contamination (11%)	Design — ventilation/ distribution (75%)	Poor filtration — low filter efficiency (57%)
	Microbiological contamination (5%)	— filtration (65%) — accessibility/ drainage (60%)	— poor design (44%) — poor installation (13%)
	Building fabric contamination (3%)	Contaminants (60%) — chemical — thermal — biological	Contaminated systems — excessively dirty duct work (38%) — condensate trays (63%) — humidifiers (16%)

- Malfunction of system due to blocked or restricted make-up air louvers;

- Temperature and humidity extremes or fluctuations (sometimes caused by poor air distribution or faulty thermostats);

- Barriers to air flow and circulation from divider walls and working station partitions;

- Improper or inadequate maintenance to the building ventilation system; and

- Inappropriate energy conservation measures: reducing infiltration and exfiltration; lowering thermostats or economizer cycles in winter, raising them in summer; eliminating humidification or dehu-

midification systems; and early afternoon shut-down and delayed
morning start-up of the ventilation system.

Inside Contamination Sources
*   Copying machines and printers: methyl alcohol from spirit dupli-
    cators, butyl methacrylate from signature machines, ammonia and
    acetic acid from blueprint copiers; submicron particles from toner;
    ozone from electrostatic copiers;

*   Unapproved (unsafe) or improperly/untimely applied pesticides;

*   Volatile Organic Compounds (VOC) that out-gas from building
    products, cleaning agents, furnishings, and finishes;

*   Boiler water additives, such as dimethyl ethanolamine, which is
    odorous and can cause dermatitis;

*   Improperly applied cleaning agents, such as rug shampoo;

*   Tobacco smoke of all types, if allowed at sites within the condi-
    tioned space;

*   Combustion gases from open flame combustion sources from caf-
    eterias and laboratories; and

*   Cross-contamination from poorly ventilated sources leaking into
    other air handling zones.

Outside Contamination Sources
*   Motor vehicle exhaust, particularly adjacent from parking garages;

*   Boiler exhaust gases, stand-by generators;

*   Re-entrainment of previously exhausted air caused by improperly
    located exhaust and intake vents or periodic changes in wind con-
    ditions from own or neighboring buildings;

*   Construction or renovation projects including asphalt, solvents,
    and dusts; and

- Gasoline fumes infiltrating basements and/or sewage systems from ruptured underground tanks.

- Nearby sewage treatment or odor producing operations.

Microbiological Contamination Sources
- Fungal contamination can occur in building components that have been flooded or wetted from condensate, such as ceiling tile, gypsum board, and insulation.

- Contamination in the ventilation system from bacteria, fungi, protozoa and microbial products; and

- Microbiological contamination commonly resulting from water damage to carpets or furnishings, or standing water in ventilation system components.

Building Fabric Contamination Sources
- Formaldehyde off-gassing from particle board, plywood, and some glues and adhesives commonly used during construction;

- Building materials and products;

- Fibrous glass insulation erosion causing dermatitis;

- Volatile organic solvents from glues, adhesives; varnishes and paints; and

- Acetic acid used as a curing agent in silicone caulking.

Findings and experience from the author and other investigators can be converted to questions to guide in-house inspections, including:

Is the air flow restricted? Are the diffusers open and unobstructed? Are they adjusted to avoid drafts? Is the exhaust system operating correctly? Are the air intakes unobstructed and operating correctly? Are air intakes bringing the building's own exhaust back into the facility?

Are the filters accessible and properly serviced? Do the filters fit the opening? Are they sealed and retained properly?

Are the coils and ductwork clean? Are make-up dampers function-ing properly during occupied hours? Are belts and baffles func-tioning properly?

Are rust inhibitors; i.e., volatile amines, getting into the airstream? Is there standing water from the humidifiers? Or is humidifier moisture getting into nearby duct work?

Has the office layout changed so that the ventilation design no longer meets occupant needs?

Have inappropriate energy conservation measures increased in-door air problems?

Is water standing in the air pathway, or do drain pans have the proper incline that allows for continuous drainage?

Is damp insulation providing a breeding ground for microbial contaminants?

Such questions need to be asked if in-house staff is to find sources of indoor air pollution rooted in poor maintenance and improperly op-erating equipment. To help guide such an investigation, these questions and others have been incorporated in the form presented in Appendix D. The Preliminary Assessment Form provides space for a summary of the interview information as well as building audit data.

Using the Preliminary Assessment Form

The forms should be used when the cause of the problem is not immediately apparent to the investigator. They are designed to help in-house personnel immediately investigate and solve indoor air qual-ity problems with their own resources wherever possible. They can, of course, be modified to fit local needs.

Those who are familiar with the health effects-source control rela-tionship will find the forms more useful and are more apt to spot prob-lems as they gather the data. Sharing appropriate sections of this book with the personnel conducting the assessment will be of benefit to them. The chapter on HVAC systems will also help in-house investigators be-come more sensitive to problems which may originate in the system.

The forms are not intended to guide in-depth inspections. Rather, they are designed to assure that the preliminary review does not miss some vital factor that in-house staff could treat. They will also help assemble the data that may be needed later by outside consultants.

At the end of Phase I, most in-house teams will find some HVAC or general maintenance items that are contributing to the problem. These are maintenance items that should be done anyway.

While the situation cannot help but benefit from rectifying these maintenance oversights, the IAQ problems may persist. While the natives get progressively restless and irritated, there is a great temptation to go beyond the "Take two aspirin..." and start "tweaking" the system. Trying to solve one problem, the uninformed, even with the best of intentions, are apt to create others. Diagnostic teams report that they repeatedly find "a maze of maintenance misapplication" that comes from these good intentions. Sound routine maintenance is warranted, but tweaking or quick fixes should be resisted. Such "Band-Aid" approaches have a way of making the situation worse.

Data Preparation

Should IAQ problems still persist "in the morning," then a more extensive examination is warranted. Unless the in-house staff has achieved "paramedical" stature, it is time to shift into the "What to do until the doctor comes" mode of operation. At this point, the data on the Preliminary Assessment Form should be verified. Any maintenance measures taken in response to the Phase I investigation as well as any observed results should be noted.

The selected diagnostic team may request additional advance data depending on the nature of the problem and a review of the preliminary findings.

Phase II: The Qualitative Walk Through Inspection

The Qualitative Walk Through (WT) Inspection may be conducted by in-house personnel if they have had the requisite training. Unfortunately, the limited availability of IAQ training for facilities managers, maintenance engineers, etc. continues to constrain the staff's ability to meet their own needs. More frequently, the WT Inspection is conducted as the initial activity of a diagnostic team.

Even if the staff is adequately trained, the emotional climate and the perceived seriousness of the IAQ problem may warrant the imme-

diate support of outside consultants. This is particularly true if more
elaborate diagnostics is indicated. The diagnostics team will need to
redo the WT to establish the appropriate air testing protocols.

In contrast to the Preliminary Assessment, the WT Inspection
involves a more thorough and detailed visual examination of the
facility and the HVAC system. It usually involves the use of simple
temperature, humidity and air flow measurement and monitoring,
humidity gauges, and a smoke pencil to check air flow. CO_2 measurements are sometimes taken to assess the effectiveness of the ventilation system.

Depending on the nature and scope of the symptoms exhibited
and the initial facility findings, inspectors may pursue specific concerns
in greater depth. WT Inspections frequently check the same areas and
equipment considered in the Preliminary Assessment, but back it up
with measurements and more expertise. WTs are designed to verify
Preliminary Assessment findings as discussed in the previous pages and
address such concerns as detailed below:

> Changes in building use; e.g., retail to office space, hospital to doc-
> tors' offices, and their implications for indoor air quality.

> Milestone activities, such as redecorating or tenant build-out.

> Effectiveness of the ventilation system. Deterioration or blockage
> of return and supply paths.

> Smoking policies and compliance behavior.

> External pollution/contaminant sources.

> Changed thermal and contaminant loads due to new furnishings,
> equipment or increased occupancy patterns.

> Inappropriate energy management measures, such as closed sup-
> ply louvers and air handling units turned off during the day or
> within certain temperature ranges. Possible modifications to VAV
> systems that could improve indoor air quality.

> Visible water leaks, condensate deposition, or water damage.

Filter efficiency and performance. Static pressure gauges as indicators. The need for additional filters downstream from contaminated heat exchange components; i.e., plenums and duct work.

Types of humidifiers (recirculating or independent steam humidification and condition; filter plate type) general cleaning schedules. Air humidification levels, especially in winter to meet the recommended relative humidity of 30-60 percent.

Measurements of the degree of circulation. Measurements of supply and exhaust flows. Ventilation systems out of balance.

Level of custodial care, maintenance; access to equipment.

Inadequate temperature control strategies, including inadequate recovery time after night or weekend set back.

Cleaning procedures (carpets) that leave irritating residues.

Surfaces and HVAC components, where microbial growth is evidenced, such as cooling coils and drain pans.

Appropriate procedures for cleaning, disinfecting or using proprietary biocides—making sure that cleaners, especially biocides, are removed before AHUs are reactivated.

Job pressures, ergonomics, occupant density, management-labor stress and ways these pressures may exacerbate work area conditions that would otherwise be tolerable.

Phase III: Simple Quantitative Sampling and Assessment Techniques
The core of the learning from this early phase of the investigation comes from the experienced and knowledgeable walk through. Many times it is inappropriate or inconvenient to perform even simple sampling and assessment procedures if not enough information is known about the problem site. However, if the opportunity and the sampling equipment is available, some basic evaluation techniques will add to the value of the WT. The following characteristics of the building can be easily and economically diagnosed and analyzed during this stage. The

resulting data aids in the understanding of the general performance of the building and its various systems. Along with the basic walk-through (WT), it also provides a foundation and direction for more elaborate and more costly diagnostics if they become necessary.

Temperature and Humidity

Ambient or operative temperatures should be checked. The ASHRAE published guidelines are intended to achieve thermal conditions, which are "comfortable" for 80 percent of the occupants in a given environment. ASHRAE Standard 55-2010. *Thermal Environmental Conditions for Human Occupancy*, temperature recommendations consider the level of occupant activity, clothing, relative humidity (RH). For example, 80 percent of the occupants in a typical office in summer with 50 percent RH should be comfortable in the range of 73° to 79°F. If the operating temperature is outside this range, then more than 20 percent of the healthy people are apt to experience some discomfort. Similarly, winter readings at 30 percent RH should fall between 68.5° to 75°F.

When feasible, temperature checks should be taken during the conditioning season; i.e., for the heating season while the boiler is on and the system is fully operational. Efforts should be made to avoid days with extremely high or low seasonal variations.

Local thermal discomfort should be checked. The vertical temperature difference should not be greater than 5°F. from head to toe, approximately 67 inches to 4 inches.

Relative humidity (RH) has long been a concern in very humid climates. More recently, the health implications for low humidity levels have received greater attention. RH below 20 percent is now associated with increased discomfort, drying of the mucous membranes and the frequency of colds. Excessive humidity, above 60 percent, can foster contamination through moisture deposition and subsequent fungal growth. (See the RH discussion in Chapter 6.)

Taken in combination with temperature, the "comfort zone" is considered to be 30 to 60 percent RH with temperatures between 68 and 76°F. (23° to 25°C).

Sampling Techniques for Temperature and Humidity

Measurements should be taken in a number of locations and times of day, particularly where workers complain that the area is too hot or too cold. Ambient temperature data from thermostats should not be accepted unless the investigator is satisfied that they have calibrated and verified recently. Currently, there are a variety of portable and hand-held instruments ideal for both spot checking and/or long term monitoring of temperature, humidity, and other IAQ parameters. Many are battery powered for versatile field usage and some have extensive electronic memory and software that enables them to acquire, store, and download data into PC computers. Prices vary dependent upon their features and complexity or capacity of their electronics and memory.

Air Flow and Outside Air

Outdoor air quantities introduced to commercial building for ventilation and dilution purposes is normally set by local code authorities. Since the 70's, these quantities have been heavily influenced by ASHRAE Standard 62 and its ongoing revisions. The investigator should be familiar with these prescribed air volumes along with their prevailing application guidance. Since the codes or applicable standards vary over time, the professionals conducting the assessment should be cognizant about the outdoor air requirements that were or are currently applicable to the building being investigated, dependent upon the timing of the construction/renovation schedule. Original design specifications, submittal data, and test and balance reports are good resources for original intent. In this relatively simple assessment walkthrough phase, the primary task is to observe the air pathway to assure that the ventilation system is functioning according to the original design intent. This should reveal any obvious malfunction of the system, whether caused by system degradation or operational intent. Issues such as blocked make-up air louvers; damper malfunction; air pathway blockages; and failure of linkages, actuators, or controls that impair the introduction of appropriate ventilation air can be identified for appropriate action. Simple evaluation tools, such as smoke pencils can be used to verify airflow and directionality. Other hand-held tools, such as the hot-wire anemometer, can be used to estimate airflow. More precise ventilation air quantification may be required in the Complex Assessment, which can be performed using TAB equipment or employing more precise CO_2 or temperature measurements to perform mass balance determination, as discussed later in this chapter.

Investigators should be aware that the *amount* of outside air may not be indicative of the level reaching occupants. The direction and velocity of air flow and it effectiveness in reaching the occupants are critical. Care should be taken with smoke tubes, especially in health care facilities, because their fumes are pungent and irritating if directly inhaled. In facilities of this type, however, dry ice is usually prevalent. When it is available, the fumes from dry ice in warm water will provide a benign but visible substitute for the more offensive smoke pencil. Be aware that the fumes are high concentration carbon dioxide, which is heavier that air. Also, the fumes will affect carbon dioxide monitoring readings if measuring is concurrent in timing and locale.

Measuring Air Flow

At this stage of the investigation, air flow is measured to be sure vents are functioning properly and the airflow is suitably directed. Exact measurements are not critical. Air movement from vents can be checked with smoke tubes, which can be obtained from safety supply houses.

Table 4-2 summarizes the temperature, humidity and air flow guidelines for occupant comfort.

Carbon Dioxide (CO_2)

Carbon dioxide (CO_2) is exhaled by the occupants in a room. Unless concentration reach exceptionally high levels, such as in excess of 5000 ppm, CO_2 is not considered a contaminant. It can, however, act as an excellent surrogate for indicating elevated concentrations of contaminants more difficult to measure.

Outdoor ambient concentration of carbon dioxide are usually in the 300 to 450 parts per million (ppm) range. The elevated levels of carbon dioxide above the 300 ppm content of normal ambient air is usually found in urban areas during peak rush hours due to automobile exhaust pollution. Experience has shown that occupants are apt to experience stuffiness, headaches, fatigue, eye and respiratory tract irritation

Table 4-2. Performance Criteria for Maintaining Thermal Environment Comfort Conditions

PARAMETER	GUIDELINES
Operative temperature (summer)	73 to 79°F (at 50% RH)
Operative temperature (winter)	68 to 75°F (30% humidity)
Air movement	$\leq$ 30 FPM;
Vertical temperature gradient	Not > 5°F between 4" and 67"

*Source—ASHRAE Standard 55

if indoor levels exceed the outside level by 2-3 times. This concentration itself is not responsible for the complaints. Rather, it is an indicator that other contaminants in the building may have increased to undesirable levels.

CO_2 is used to define a minimum limit of outdoor air needed to ventilate a building since it is a reflection of the number of human occupants and their metabolic rate. Thus, CO_2 levels are a useful surrogate for diagnostics purposes. CO_2 above this point is not necessarily indicative that the building is hazardous or that it should be vacated. It just sends a signal that the ventilation is probably inadequate to dilute occupancy derived odors. Persily, *et al.* in a NIST document, provided insight and guidance for CO_2 monitoring practices which has been developed into an ASTM standard # D6245-02. His work emphasizes the importance of using the outdoor air as a reference point. Because of the potential elevation and fluctuation of outdoor levels, a differential of 700 ppm over the outdoor level is suggested as a more appropriate target than the obsolete designation of indoor level of 1000 ppm.

Sampling Techniques for Carbon Dioxide

Historically, detector tubes, which indicate CO_2 concentration by the length of color change on a sampling tube, provided an easy means of determining CO_2 levels. However, the sampling tube is highly erratic in accuracy and is highly subjective in interpretation. The technique requires the sample to be manually acquired in close proximity to the investigator, which can introduce substantial operator source induced errors. Further, the tube technique allows only single grab sampling that makes it difficult to profile the operating cycle of the building. More recently developed portable or hand-held instrumentation provides more accurate and reproducible data. They also allow for remote and long term monitoring which provides operator error-free data. The memory capacities and computerized data management enable the long-term data to track and profile the building characteristics over a variety of responses to varying occupancy and operational tactics.

Sampling locations should be planned ahead and should include areas where problems are suspected and *areas where they are not*. If no areas have been specifically identified as problem locations, then sampling locations should be selected from each area of the building. Information from a range of locations provides valuable data on the distribution of the air and how the various facility systems respond during normal occupancy.

CO$_2$ readings should be recorded by the specific location and the time of day. The pattern by locality can provide skilled investigators valuable information as to flawed ventilation effectiveness.

Interpreting Results from Simple Tests

Nothing more than common sense is required to interpret and act on some of these findings. If temperature, humidity or air flow readings compare unfavorably with the guidelines, then the system should be adjusted to bring the conditioned space into compliance with design intent or prior performance.

CO$_2$ readings require more interpretation, but some crude guidelines can prove helpful. If indoor and outdoor CO$_2$ readings maintain the same relationship and remain that way throughout the day, chances are the amount of outside air entering the facility is adequate. The problem, instead, may come from an imbalance in the ventilation system or with the temperature/humidity.

If CO$_2$ levels are about the same as those outdoors in the morning but rise during the day, the air exchange rate is probably all right but not sufficient for the current level of occupancy or the activities taking place during the day. If the CO$_2$ level starts out unacceptably high, the system may have been shut down too early on the previous day (known in energy parlance as "coasting"). The system also may not have been turned on in the morning long enough before occupancy.

These interpretations of CO$_2$ *vis a vis* ventilation are grossly simplified using whole building averages. But knowing CO$_2$ levels for the whole building only tells part of the story. Distribution problems have been identified in up to two-thirds of the SBS buildings investigated; so it is important to look at ventilation in specific areas as well. Ventilation testing should not be limited to, and should never be reported solely as, whole building averages. Knowing the value for the whole building is useful information, but it can mask critical space-to-space or zone to zone variations. Ventilation rates within a building can vary by factors of five to ten. "Open space" offices and schools that have been partitioned are likely candidates for dead air pockets.

The importance of ventilation effectiveness, the ratio of outside air to the amount of air reaching the occupant zone, is discussed in the chapter on thermal comfort. To the occupant, it really doesn't matter how much air is entering the building if it is not reaching them. Testing ventilation rates in a variety of locations, *where the people are*, gets to the places where the complaints originate.

In using CO_2 to interpret contaminant levels, we also need to remind ourselves that CO_2 is essentially an occupant-based measurement and some contaminants are not occupant related. In a room with few people, for example, the CO_2 could be well below accepted levels while concentrations of some VOCs could still be unacceptably high.

In other instances, CO_2 as a surrogate may indicate higher levels of contaminants, but increased ventilation may not be the best or most cost effective solution. In almost all cases, removal of the containment at the source is preferable.

In other instances, either CO_2 or temperature values may be used to evaluate or approximate the amount of ventilation air that is being introduced to the indoor environment. Either of these calculations is based upon mass balance equations that are simple to solve for the ventilation fraction by substituting the measured factor. The following outline should be followed to determine adequacy:

(1)   *count* the number of occupants in an area and multiply by required cfm/person,
(2)   obtain the recorded total volume of air being recirculated from a recent balancing report or by measuring with a Pitot tube
(3)   calculate the percentage of air required

$$\% \ Outdoor \ Air = \frac{No. \ of \ people *^{cfm}/_{person} * 100}{Total \ flow \ of \ system}$$

(4)   with outdoor intakes in minimum position, measure the temperatures
    —   return temperature (RT) downstream from return fan
    —   mixing temperature (MT) at several points up stream from the supply fan; average the readings
    —   outdoor temperature (OT) in the vicinity of the intake,
    It is best to do this when there is a significant temperature difference between outdoor air and return air; i.e., $90 < OT < 60$.
(5)   calculate the outdoor air delivered by the system

$$\% \ Outdoor \ Air = \frac{RT - MT}{RT - OT} \times 100 \ and$$

(6)   compare required outdoor air to delivered amount. If values differ, adjust. Wait approximately 10 minutes and repeat steps 4, 5, and 6.

Using Outside Support Effectively

If an outside team is used, the diagnostic team will first seek to identify the nature of existing (or potential) IAQ problems, and within that context define the scope and objectives of their investigation. They will recap many of the activities, review building plans and documentation in detail, and interview key personnel.

Management should make someone available to the team to expedite their work. It takes valuable time and money for the team to fumble through files or to try to find someone who knows about a piece of equipment. An individual who knows the organization and the facility should be available to the team to gather data, coordinate meetings and interviews, and support the team's investigation. Personnel to be made available for interviews/meetings should include senior administration and health/safety officers to review administrative matters, and the facilities manager plus technical and O&M staff to review facility-related concerns. Interaction with the occupants during the team's first inspection of the facility may prove valuable and management should encourage the team to engage in such dialogue.

This diagnostic team walk through is frequently a scoping audit to develop hypotheses regarding problem causes, how they appear to be manifested and relationships to building, equipment and O&M characteristics. Simple measurements already discussed; i.e., temperature, RH, air flow and possibly CO_2, will usually be taken.

From this information, the team will generally formulate a preliminary hypothesis and a diagnostic procedure predicated on need, feasibility and cost-effectiveness.

The team usually concludes this phase with a meeting with management to review activities to date and present preliminary findings. Actions that can be taken without further diagnosis or analysis being recommended, such as O&M procedures, simple adjustments to the ventilation system, and/or changes in occupant behavior. If the team believes additional diagnosis will be needed, it may discuss the nature of that work at this time.

By the end of Phase III, approximately 80 percent of the problems causing SBS or BRI have been identified. If problems still persist, then further diagnostic procedures will be needed. From here on, however, the going gets harder and the costs start to mount.

Phase IV: Complex Quantitative Diagnosis

If this phase constitutes an outside team's introduction to the facility, it will most likely review the work done by the in-house staff and then proceed to repeat all, or almost all, of the work just described. This is necessary to provide them first hand information and experience about the building, its systems, and how it works. It also provides them with the basis for understanding the "what, where, when, and how" of more complex diagnostics. Without this kind of information, there will be tremendous waste in the sampling process.

In addition, more analytical procedures will be conducted on-site and back at the team's offices. The characterization of the symptoms, complaints and problems will usually influence other evaluation procedures. Medical evaluation may be involved. Typically, the major focus at this stage will usually be an engineering analysis of the HVAC system, the controls and the overall building performance to help establish environmental performance criteria.

Measurements may be taken of implicated pollutants. For example, if observable indicators (symptoms and building assessment data) point to carbon monoxide as a contaminant; then, instantaneous and periodic monitoring readings may be conducted using CO sensing instrumentation.

Various diagnostic teams may use other measurements or air constituents usually, as surrogates, during the simple diagnosis phase. Their selection is predicated on the type of facility and the information gleaned from the scoping audit. For example, certain tracer gases, such as toluene, can be used in an office building as a surrogate for the intensity of human activity and the related adequacy of the ventilation system.

More precise data may be needed on the ventilation system, such as ventilation efficiency. CO_2, as one means of measuring ventilation was discussed under Phase II. Ventilation can also be measured with tracer gas, such as sulfur fluoride (SF_6). Tracer gas measurements can qualitatively and quantitatively verify the need to adjust or modify the ventilation system. Using either approach, the system deficiencies are easy to evaluate and sometimes easy to remedy. Figure 4-2 offers a simple schematic of the single tracer gas technique for measuring infiltration and ventilation performance in a large building.

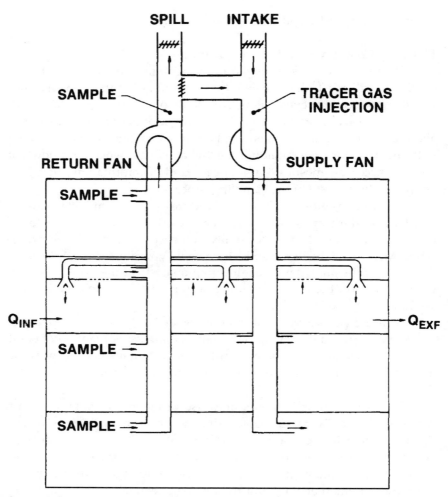

Source: U.S. Department of Energy

Figure 4.2 Single Tracer Gas Technique for Measuring Infiltration and Ventilation Performance in Large Buildings

As previously noted, however, it is important to know what is going on throughout a building, and not just rely on some over-all air exchange value. In addition, as the diagnostics become more sophisticated, correlating the simultaneous ventilation rate measurement to reported contaminant concentrations in a given area offers a much better picture of what is actually happening around the occupants in that location.

Researchers frequently use gas chromatograph (GC) and mass spectrometer (MS) tracer systems together. Chromatography, around since the 1950s, has become a good tool with the development of synthetic adsorbents suited to organic compounds and has proved useful when scientists have a rough idea of what they are looking for. A more sophisticated analysis device is the mass spectrometer, which can identify chemical compounds in the air by their atomic fragment "fingerprints." The United States Department of Energy cites the relative advantages of each system in Table 4-3.

Table 4.3. GC and MS Tracer Systems

RELATIVE ADVANTAGES

[ ] GAS CHROMATOGRAPH (GS)
    - High sensitivity—low tracer usage
    - Lower unit cost—distributed systems
    - Application: large buildings; steady-state measurements
[ ] MASS SPECTROMETER (MS)
    - Mass range 1-300 amu—wide range of compounds
    - Fast response time—active control
    - Application: small buildings; dynamic measurements

Figure 4-3 depicts a multi-gas tracer measurement system using an MS system.

This diagnostic approach may require quite sophisticated simulation techniques to reflect the real life dynamic of building performance as well as tests for implicated pollutants or as surrogates. It requires a diagnostic team with broad expertise. From the management perspective, it is important that the team offer a multidisciplinary approach in its diagnostic procedures. This is discussed in greater detail later in the chapter under "Selecting a Diagnostic Team."

The results from this phase often finds diagnostic team recommendations indicating that the original HVAC design was inadequate, or the system can no longer keep pace with changes in the facility or its use. In either case, if the design is inadequate to cope with the contaminant or thermal loads, the remedy is usually HVAC system modification. Recommendations for source control are also likely. It is not unusual,

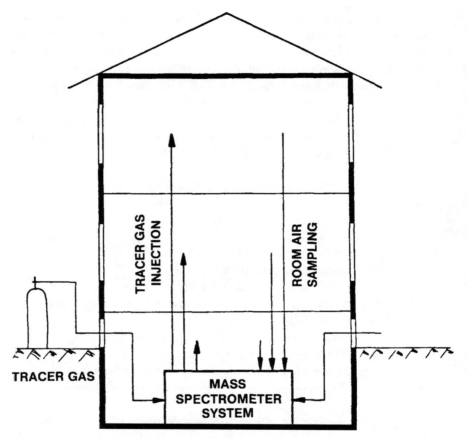

Source: U.S. Department of Energy

Figure 4-3. Multi Gas Tracer Measurement System

even at this stage, for more operations and maintenance measures to be recommended. At this point, however, it becomes critical to perform the obviated mitigation or corrective actions. No further time, money, effort, productivity, or human relations should be sacrificed to "further analysis" until the obvious is repaired or brought up to expectation.

If the implemented recommendations from prior phases do not resolve the problem, then more complex diagnostic steps are required. Or, elaborate and detailed measurements and analyses may also be required to respond to litigation proceedings. Occasionally, they are needed to assess conformance to certain performance criteria.

These tactics can be time consuming, costly and complicated. It should not be undertaken until other alternatives have been exhausted. During this phase, objective measurements of chemical, physical and microbiological parameters are paired with subjective responses from the occupants to their environment. Specific tests and real time measurements may be needed in a range of locations.

These locations usually include sites of complaints, areas with susceptible occupants, areas where no complaints have been registered, and an outdoor or remote control site. Other areas that may be tested may include offices connected to the complaint area by the ventilation system, or areas with varying levels of ventilation efficiency.

Management should ask for and expect explanations in lay terms regarding analytical procedures to be used and procedures for selecting and applying appropriate sampling methods. The team should also be requested to describe methods it intends to use to provide quality assurance/quality control that will assure an acceptable level of accuracy, precision and reliability. The organization's spokesman and people responsible for internal communications should be fully apprised of actions to be taken and be able to explain them accurately in lay terms.

Recommendations from the complex diagnostics may call for a range of mitigating actions including, source control, system modifications, and maintenance and operational procedural changes. Again, no further monitoring or assessment should be performed until the already obviously indicated mitigation has taken place. Should more elaborate diagnostics be indicated, the following discussion provides an overview of the more common advanced building diagnostic practices. The intent of the section is to provide understanding and working knowledge, but not be an instruction manual. The facility manager should note at the outset of this discussion that building evaluation and diagnostics is still an emerging art form. Further, the interpretation of the results must rely on the broad multi-disciplinary experience of the investigator. Lastly, the process should not be entered into without the awareness that it is an expensive, time consuming, and sometimes frustrating process. Adding to the cost and difficulty of analysis is the requirement for sufficient sampling sites to avoid false positives and negatives, as well as the need for control sites for comparison. This level of evaluation should never be undertaken without performing the prior phases. This can happen in response to the loud and passionate request that commences "Come test my air! I think I am sick from bad air!" As stated in the Jan. '97 En-

gineered Systems IAQ Report "Any IAQ evaluation that conducts extensive air sampling without thorough investigation of the HVAC system is,
at best, providing only a superficial view of the IAQ status of a facility."

Carbon Monoxide

If carbon monoxide is suspected, specialized sampler tubes can
be used for quantification. However, the preferable method is to use a
measuring instrument equipped with specialized CO sensing probes.
This technique enables longer term monitoring to both quantify the concentrations, but also track the timing of peaks throughout the building
cycle, in order to evaluate concurrent potential generation activity. CO
seldom reaches actionable limits in commercial buildings, but can be an
effective tracer of combustion sources. It also can be a useful surrogate
for identifying high pollution periods in the outdoor air.

Particulate Matter

If high levels of particulate matter are indicated, it can be evaluated as either "mass" or bulk content or specific particle counts of
specific sizes. The mass data are helpful for comparative quantification
because most external environmental data are noted in mass, such as
the NAAQS information from USEPA. This allows direct comparison to
known outdoor and action levels. The data is developed using a calibrated air sampler and a prescribed collection filter cassette. Dependent
upon methodology, data can be derived for specific size cuts, such as
PM 10 or PM 2.5.

Particle count data can be more helpful in diagnosing indoor air because it enables a more precise analysis and comparison of particle size.
This can be helpful in tracking sources as well as determining specific
mitigation tactics, such as appropriate filter selection. The use of particle
counters also enables the field evaluation of the efficiency performance
of installed filtration systems. These data are developed using an instrument that discreetly sizes and counts individual particles carried in the
air stream over a size range from .3 microns to 10 microns. The better
units have multiple size channels and are equipped with memory capability enabling down loading and data manipulation. A more recent
technology enables the recognition of superfine (meaning sub-micron
sized) particles and is embodied in a hand held instrument similar to a
Geiger counter that will trace sources of ultrafine particles. The tracking
of ultrafine particles can help understand contaminant pathways and

infiltration, since outdoor air is a primary source of ultrafine particles. It also can assist in the on site tracking of fungal activity in concealed sources of mold spores, if the fungal spores are the primary source of the fine particles. Note the following discussion regarding microbial growth for further information on fungal determination.

Microbiological
    Often, visual evidence of mold is sufficient to warrant mitigation and clean-up of fungal growth. In fact, visible growth prevails over air sampling as evidence according to leading microbiologists. However, further evaluation may be required to qualify and speciate the type of microbial matter or to quantify it for comparison and verification of clean-up tactics. Although, both bacteria and fungi can be sampled using these techniques, the predominance of evaluation in commercial buildings is based upon fungal analysis. Since most fungi spores and other airborne components are small, they can be demonstrated using particle counters. However, this devise cannot discriminate the fungal particles from other nonviable matter. Thus, more specific and selective techniques are employed using microbiological laboratory protocols. Passive sampling techniques include bulk sampling and surface sampling. Using either a bulk sample of the surface material, or by using sticky tape, or surface swabs to sample the activity on the surface, the evidence of growth can either be observed by microscope, or the viable spores can be cultured and grown into colonies and identified. This technique provides the opportunity for affirmation of growth and identification of genus and sometimes species; however, care should be taken in interpreting the results because even a normal surface may have relatively high counts of settled spores. The resulting high "Oh My God" numbers may cause undue concern by occupants when, in fact, they may be meaningless.
    Active sampling of airborne microbial contaminants is performed using a calibrated pump which pulls a known volume of air over a glass slide, nutrient agar, or filter. When collected onto a glass slide, the sample is typically analyzed microscopically and reported as spores/cubic meter. This sampling is commonly referred to as "spore trap" and has the advantage of enumerating both viable and non-viable spores.
    When the air is collected onto a nutrient agar, the laboratory will incubate the sample to grow colonies. The resulting colonies are enumerated and reported as colony forming units (CFU)/cubic meter, The

sampling is commonly referred to as viable or culturable sampling and has the advantage of possible identification down to the species level.

Samples collected on the filter can be analyzed in the lab via microscope, culture, or more recently, via polymerase chain reaction (PCR).

Because of the ubiquitous nature of mold spores, outdoor air must always be sampled as a control. Counts that are no more than outdoor air and consisting of similar speciation are considered normal regardless of the specific count. Indoor readings that exceed outdoor levels, or indicate species not in the outdoor profile are indications of interior sources or amplification. Because of the concern for specific fungi as agents of potentially serious disease and health effects, care should be taken whenever the more toxigenic species are reported. Because of involvement with infant deaths and implication in other highly publicized IAQ incidents, *Stachybotys chartarum* is of particular note. The latter fungus yields rather large and relatively "sticky" spores and is, thus, seldom found in airborne samples. Therefore, when it is found in airborne samples, special concern is warranted. Although Stachybotys started the frenzy about "black mold" which was quickly labeled "Toxic Mold" by the media, there are other species that produce Mycotoxins that can also induce adverse symptoms from exposure, such as species of the genus Aspergillus. All airborne fungal matter should be considered a contaminant of concern whenever high concentrations or extended exposure periods are experienced by occupants. During active growth, fungi also give off MVOC (Microbial Volatile Organic Compounds) that are, in some cases, specific to species and can represent a distinctive "footprint." This can be traceable using precise air capture and evaluation techniques and highly specialized laboratories. In a general text such as this book, there is little purpose in going into great detail regarding specific fungal species, their properties, and their health effects. However, there are several excellent web-based data sources that provide long glossaries of environmental fungi. One such site is U. of Minnesota Environmental Health and Safety Dept. @ *www.dehs.umn. edu/iaq/fungus/glycoglos*. Fungal data resulting from indoor airborne sampling should always be considered in the context of outdoor data and related visual, odor, and bulk sampling.

Bacteria, such as *Legionnella* are found predominantly in reservoirs of aged or stagnant water. Thus, they are best monitored through laboratory evaluation of bulk water samples acquired from the pooled water source.

Volatile Organic Compounds

The precise and cost effective sampling of VOCs is probably the most difficult of the diagnostics techniques. Much of the early misinformation derived from IAQ analysis came from the VOC area. This is because much of the early data acquisition of this type data was done using industrial hygiene techniques that are designed for much higher concentrations. Thus, there were many reports ending "You have no problem—you do not exceed any OSHA industrial guidelines." This, of course, flew in the face of ongoing and continuing occupant complaints. It is now known that most IAQ complaints regarding VOCs are well below established industrial limits. Air capture, sampling techniques, sample protection, and evaluation methodologies are complex and require highly competent laboratories specializing in GC/MS equipment, training, and competency. Using calibrated air pumps and long sampling times, air is drawn through sorption tubes filled with selected sorbents. The tubes are then thermally de-sorbed and evaluated using GC/MS techniques and elaborate computerized databases for chemical identification and quantification. Odorous chemicals can be evaluated similarly but may require longer sampling periods, special sorbent tubes, and much more experience and coordination with the analytical laboratory staff in the identification and quantification process.

An alternate method of collecting air for analysis is through the use of chemically inert, stainless steel cannisters that are pressurized with a sampling pump. The disadvantages of cannisters are their size and handling limitations as well as potential for crossover of contaminants that may adsorb onto internal surface. In the case of higher concentration VOCs, a number of devices employing photo-ionization and flame-ionization (PIDs and FIDs) are available as hand held portable devices that can be employed for screening. Previously, these devices were not sensitive enough for IAQ level concentrations, but more recent technologies are refining their threshold sensitivities.

Formaldehyde

Although formaldehyde and its aldehyde chemical family members are defined as volatile and organic, they cannot be evaluated using the routine VOC test. HCHO is so volatile and reactive, that it breaks down during the thermal de-sorbing process and is lost in the sample. Because it is so ubiquitous, and because of a long concern by health officials about its potential adverse health effects, a number of evaluation

Table 4-4. Advantages and Disadvantages of
Common Formaldehyde Sampling Methods for Indoor Air

Method	Collection Media	Analysis	Sensitivity (ppm) 15 Min	Sensitivity (ppm) Long Term	Advantages	Disadvantages
Detector tube	Reactive Adsorbent	Visual Color Change	0.5	NA	Easy Low Cost	Accuracy problems Affected by humidity and temperature extremes Not sensitive enough
Passive Monitors	Adsorbent	Spectrophotometry	0.8–10	0.05–0.3	Easy sample collection and analysis	Sensitive to humidity and air velocity Poor sensitivity for peak measurements
OSHA Method 52	Adsorbent tube (XAD)	Gas chromatography	0.02	0.02 (2 hrs) (at recommended sample volumes)	Good sensitivity Stability	Media obtained from lab Special analytical equipment Time lag between sampling and analytical results
Chromotropic Acid—NIOSH P&CAM 125	Impinger	Spectrophotometry	0.16	0.04 (1 hr)	Sensitive Selective Simple	Interferences Time lag
Pararosaniline	Impinger	Spectrophotometry	0.02	0.005 (8 hr)	High sensitivity Simple	Impinger solution is toxic Interferences Time lag
Real-Time Monitor	Continuous	Infrared	0.4	NA (Instantaneous)	Easy Fast response Real time	Interferences Poor sensitivity at low concentrations

methods exist. The most widely used method is the passive sampler. However, the impinger, which is based upon wet chemistry, provides more accurate data. The comparison in Table 4-4 will provide additional information on the various methods and apparatus.

Odors

Quantitative evaluation of odors is possible using techniques that are similar to routine VOC testing. However, often the concentration thresholds are reduced even further. It is also helpful to be working with a laboratory that specialized in odor analysis as it takes special technician and data base expertise. It is also helpful if there is some working field knowledge that provides early insight as to the potential or probable nature of the odor. The better way of saying this is that the evaluation is easier if the investigator already knows the answer, but is looking for confirmation.

MVOC (Identification of Microbial Growth

This evaluation process recognizes the nature of microbial growth that gives off a wide range of organic gasses as part of the life chemistry. It is proposed that specific species give off significant and distinctive chemical compounds that can be used as a diagnostic footprint called Microbial Volatile Organic Compounds (MVOC). Although this area is a field of active research, there is promise that specific compounds such as ergostol and certain ketones and alcohols will provide a dependable gas phase identifier that will be more accurate than the vagaries of microbial culturing. The technique will also provide helpful insight when no other visible or viable clues are fruitful.

Phase V: Proactive Monitoring and Recurrence Prevention

Once an indoor air problem is resolved and the complaints are only a faint memory, the job is not over. One has only to recall the hysteria and harangues, not to mention the lost productivity and the investigation costs, to resolve not to get caught in such a bind again. An IAQ program, with an integral preventive maintenance program, addressed elsewhere in this book, helps to avoid such a dilemma.

In addition, monitoring procedures through routine checks and periodic measurements can help assure the program and the remedial measures are working.

"Monitoring" is sometimes used in IAQ literature in relation to investigative procedures, measurements and equipment use, such as radon monitors. When used in that sense, monitoring denotes activities prior to mitigation. Phase V monitoring is proactive and is used in the broader, more conventional, sense of assuring informal and continuing compliance to some measure of acceptability following remedial action. Phase V monitoring is designed to check up on and keep track of measures taken to be sure beneficial results are preserved. (See Figure 4-5 for illustration.) The benefits of proactive monitoring are outlined below:

• Provides baseline for comparison with prior or future building performance parameters;

• Provides early warning of drifting or erosion and degradation of building performance to trigger proactive preventative maintenance;

• Assures the owner, occupants, tenants, and management team of the on-going air quality of the facility;

PROACTIVE MONITORING: TVOC PROFILE

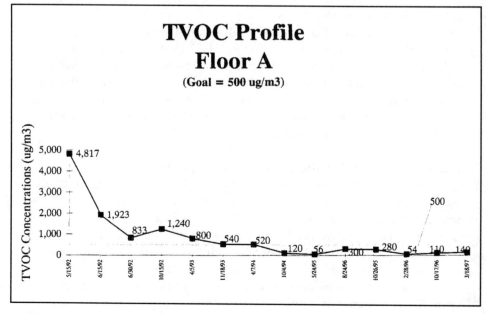

Figure 4-5. Monitoring Data Demonstrate Reduction of TVOC Concentration Over Time (Source: Milam AIA Healthy Buildings Symposium)

- Provides evidence of proactive rather than negligent IAQ posture in case the building air quality is challenged for any reason;

- Provides evidence of building IAQ performance in case the building air quality is challenged by an occupant;

- Provides investigators data of trends and historical performance of the facility to aid in diagnostics should an IAQ incident occur;

- Provides documentation for marketing high quality space to tenants and prospects; and

- Provides documentation for due diligence during property ownership changes.

MEASURES OF ACCEPTABILITY

At this point, it helps to know just a little about acceptable levels, how pollutants are measured and the problems associated with measuring contaminants.

From the safe falling from a cliff toward the roadrunner in the cartoons... to the fuselage explosion due to a faulty cargo hold latch... to a particulate that migrates into the deepest reaches of an employee's lung... all can be lethal. All these scenarios involve increasingly smaller objects and increasingly more difficult sources to identify. We are left with no doubt where the ACME safe came from or when. In the case of an airborne contaminant, however, we may not know when exposure occurred, the length of that exposure or the contaminant concentration.

Furthermore, we know the cartoon character, who was squashed flat in the last frame, will be up devising new diabolical schemes in the next. In contrast, pinpointing the extent of damage from a particulate or even knowing if the damage was caused by an airborne contaminant is difficult, sometimes impossible.

The health effects may not become apparent until long after exposure, as in the case of asbestosis. IAQ measurement is a subtle science made more difficult by many factors operating at once, including the exposure times, frequency and dose. To complicate things further, a particulate, for example, may have picked up a passenger or two, like radon daughters or VOCs. Intervening and confounding variables may make the identification of single contaminant very difficult and its source nearly impossible.

To make matters worse, there is a lot we still don't know about acceptable levels. And now that we are to obtain more reliable measurements, we aren't always sure what we've found out. Some data are available on acute exposure to certain pollutants in an industrial setting. Far less may be known about chronic exposure at lower levels in an office building. Our limited understanding of the damage some contaminants can do at low concentrations over a long, long period of time is the reason elimination at the source is much preferred to control by ventilation.

Other factors, such as age, general health or sensitivity, and prior exposures, may play a greater role for some individuals than dosage or exposure duration. In many situations, children and the elderly are more vulnerable to a contaminant of concern.

In many instances, procedures for measuring a contaminant have been limited to research settings. Some techniques/equipment are very expensive. Few people are well-grounded in effective measurement procedures. Once data are obtained, only a limited number of analysts can accurately interpret those findings.

In this milieu of uncertainty, establishing measures of acceptability and knowing with confidence that they are being met is hardly an exact science. Nevertheless, acceptable levels can prove helpful. Furthermore, extensive research in the engineering, medical, and psychosocial sciences is underway. This will do much to push the envelope of technology of assessing cause and health effect, exposure limits, and successful prevention and mitigation tactics. Limited research is also underway to examine the existing building inventory to assess the air quality factors relating to productivity in "sick" and "healthy" buildings. EPA has been underway since 1994 with a building study called "BASE" (Building Assessment Survey and Evaluation). Some contaminants and currently accepted ambient air quality guidelines/standards depicted in their usual units of measure are shown in Table 4-5.

In addition to the noted hard targets, it is worthy to note several softer targets that are promulgated by various experts. In the area of microbials, fungal counts in the 200-500 cfu (colony forming units)/cubic meter range are considered normal if this falls below outdoor levels and involves similar species populations. However, if outdoor levels are much lower or show different population make-up, then amplification may be indicated. In VOCs, counts in the 200-500 microgram/cubic me-

Table 4-5. National Ambient Air Quality Standards.

Pollutant	Primary Stds.	Averaging Times	Secondary Stds.
Carbon Monoxide	9 ppm (10 mg/m^3)	8-hour[1]	None
	35 ppm (40 mg/m^3)	1-hour[1]	None
Lead	1.5 µg/m^3	Quarterly Average	Same as Primary
Nitrogen Dioxide	0.053 ppm (100 µg/m^3)	Annual (Arithmetic Mean)	Same as Primary
Particulate Matter (PM$_{10}$)	Revoked[2]	Annual[2] (Arith. Mean)	Revoked[2]
	150 µg/m^3	24-hour[3]	Same as Primary
Particulate Matter (PM$_{2.5}$)	15.0 µg/m^3	Annual[4] (Arith. Mean)	Same as Primary
	35 µg/m^3	24-hour[5]	Same as Primary
Ozone	0.075 ppm	8-hour[6]	Same as Primary
	0.12 ppm	1-hour[7] (Applies only in limited areas)	Same as Primary
Sulfur Oxides	0.03 ppm	Annual (Arith. Mean)	-------
	0.14 ppm	24-hour[1]	-------
	-------	3-hour[1]	0.5 ppm (1300 µg/m^3)

[1] Not to be exceeded more than once per year.

[2] Due to a lack of evidence linking health problems to long-term exposure to coarse particle pollution, the agency revoked the annual PM$_{10}$ standard in 2006 (effective December 17, 2006).

[3] Not to be exceeded more than once per year on average over 3 years.

[4] To attain this standard, the 3-year average of the weighted annual mean PM$_{2.5}$ concentrations from single or multiple community-oriented monitors must not exceed 15.0 µg/m^3.

[5] To attain this standard, the 3-year average of the 98th percentile of 24-hour concentrations at each population-oriented monitor within an area must not exceed 35 µg/m^3 (effective December 17, 2006).

[6] To attain this standard, the 3-year average of the fourth-highest daily maximum 8-hour average ozone concentrations measured at each monitor within an area over each year must not exceed 0.08 ppm.

[7] (a) The standard is attained when the expected number of days per calendar year with maximum hourly average concentrations above 0.12 ppm is ≤ 1, as determined by appendix H.

(b) As of June 15, 2005 EPA revoked the 1-hour ozone standard in all areas except the fourteen 8-hour ozone nonattainment Early Action Compact (EAC) Areas.

ter range are considered to be low. Levels in excess of 2500 tend to be related to complaint situations.

INVESTIGATION DIFFICULTIES

In order to have a better understanding of the investigative process and what is doable, it helps to have some appreciation of the difficulties inherent in the process. Some of the same concerns that impinge on setting acceptable levels also intrude on investigation procedures.

Multifactorial Concerns

The contaminant burden in the built environment is both complex and dynamic. It is influenced simultaneously by multiple contaminant sources that coexist: the outdoor environment and its cyclical dynamics; the occupants and their activities; the building construction components and their out-gassing products; furnishings and finishes; and the purposeful activities performed in routine cleaning and maintenance. Conversely, the interior burden is tempered by dilution from cleaner spaces or outdoors; reduced by filtration and air cleaning systems; lowered by natural decay and deposition on surfaces and "sink" sites; and modified by chemical reaction. To compound this complexity, there is mounting evidence that chemical reactions can combine airborne contaminants, such as VOC and ozone, to form even worse compounds that are highly odorous and irritating. This phenomenon can also convert gas phase chemicals to respirable particulate solids in the form of condensation nuclei (as products of the reaction). To add to the design burden, some readily identifiable contaminant sources are both gas phase and particulate in nature. For example, smoking is no longer the major indoor air quality issue that it was prior to the middle '90s. However, if the allowed smoking break-area is outdoors, but adjacent to an outdoor air louver, the designer must be aware that this problem contaminant source is both ultra-fine particles requiring very high efficiency particulate filters, but also an odorous and irritating gaseous burden, requiring a high efficiency gas adsorption. Perhaps the most confounding is the interplay of contaminant sources and human response to environmental conditions, such as varying temperature and humidity. This can include the negative response to environmental factors usually outside the HVAC designer's area of expertise, such as lighting and sound levels. These complexities lead us to the reference of "IEQ" as a broader and more complete label than IAQ. In addition, psychological factors, including job satisfaction and stress, are thought to affect some symptoms and certainly contribute to complaints. Please refer to the ASHRAE Guideline 10 for further discussion regarding the inter-reaction of various factors influencing the indoor environment.

The Human Dimension

For an indoor air problem to exist, there must be a receptor that is unfavorably affected by an airborne agent. Humans, as receptors, do not respond uniformly. Age, gender, nutrition, susceptibility, general health, ability to adapt and thresholds of sensitivity vary among individuals. All these factors may act, separately or in combination, as confounding variables. These variables may be further influenced by the comfort expectation held by the individual.

Not only do human have different expectations and varying reactions, they will report those concerns differently. Gleaning information from occupants is not without its problems. Investigations, for instance, are almost always prompted by what is *perceived* to be a problem by somebody. It will come as no surprise to students of human nature that one complaint, or the knowledge of some exposure to a potentially harmful substance (even at very low levels), will very likely foster additional complaints. A reporting bias will usually come from what Gable *et al.* call "epidemic psychogenic illness." Separating out the effects of interactions of low levels of exposure from "hysterical" behavior is exceedingly difficult.

Knowing What's "Right"

With regard to the contaminants themselves, little is known about acceptable levels of most pollutants. Even less appears to be known about low concentrations imposed over prolonged periods of time.

In many instances, the equipment, the measurement, the interpretation of the findings, or all three are far from perfect.

While the technology of measuring indoor pollutants has leaped ahead in the past two decades, testing is still in its infancy for many contaminants. Even though we can now obtain detection levels of one part per quadrillion, testing equipment available on the market for field work is costly and/or may be too delicate, non-portable, or unreliable. In addition, talented professionals truly qualified to administer and interpret results are still in short supply.

From a more scientific perspective, any guidance for management is seriously limited by the dearth of studies about *healthy* buildings and the lack of any control buildings in most research studies. European Studies have been helpful, such as a series of investigations and surveys of Danish town halls that have provided helpful insight into the influences upon the indoor environment.

Establishing any type of baseline or standard protocol has been thwarted by a lack of data, lack of uniformity in approach and the diversity of the disciplines involved. Each discipline views the problem/solution through eyes constrained by training and experience. To borrow an old line: The carpenter sees every problem as a nail and every solution as a hammer. Likewise, the engineer makes a beeline to the HVAC system while the industrial hygienist grabs their sampler air pump. It is this diversity that makes consultant selection so critical.

SELECTING A DIAGNOSTICS TEAM

A single consultant might be retained to develop a program, design an HVAC system, brief staff, train O&M personnel or perform a myriad of other IAQ related management services that cannot be handled in-house by some organizations. But hiring a single consultant to investigate a serious IAQ problem can be a serious mistake.

Unless management has narrowed its concern to one contaminant; e.g., radon, the multifactorial nature of sick buildings requires broader input. Experience has shown that standard epidemiology and industrial hygiene evaluation techniques may be inconclusive and unconstructive.

An epidemiologist, for example, usually sees the essential starting point in any building-related outbreak of illnesses as the complete characterization of the symptoms, complaints and their patterns. The industrial hygienist, following successful industrial research patterns, is inclined to test for specific pollutants. The engineer, who wants to get to the source of the problem, is apt to set aside the human research and head for the HVAC system and building operation. The microbiologist will take numerous fungal samples and "break the budget" on lab fees before the first dollar is spent on clean-up, remediation, and moisture control.

Elements of every approach are important; and, while each has merit, the problem is frequently broader than the discipline. After costly testing by an industrial hygienist (IH), for instance, a contaminant in an office may be found to be well below accepted industrial exposure levels and the management is advised that the quality of the indoor air is acceptable (a false negative). Unfortunately, we have found that we are playing by different rules and acceptable levels for contaminants in an industrial work place do not transfer to other work locations. Or, the problem may be that other contaminants are at work and this focused approach doesn't tell the whole story. In either case, after expending time and money, the management is given a clean bill of health by the IH, but the symptoms persist. Management says, on good authority, that a problem doesn't exist. Employees think management is playing games and stonewalling the issue.

Every situation, every building is different, but selecting a diagnostic team is a fairly uniform process. The management needs to identify general task(s) that need to be done, set forth the criteria; e.g., experience, and ask for proposer's credential to meet those objectives. Experience in IAQ investigations is essential. Experience in similar facilities is desirable.

Except in general terms, it is very difficult for the administration to describe exactly what is to be done. After all, if it was possible to be that precise, the problem would already be identified. It is imperative, therefore, that the references of prospective teams be requested and that they be *checked out*.

As with other contracts, it's also important to control assignability. The annals are full of stories of administrators, who thought they were getting experienced senior personnel, only to find the work being done without supervision by the June graduate. In a relatively new field, like IAQ, where talented professionals are still sparse, accountability and experience become critical considerations.

Potential diagnostic teams should be asked to describe in lay language what they propose to do and how they propose to do it; then, they should offer estimated costs for the proposed work. It is advisable to get scope of work and costs in advance of each phase. The earlier steps are more routine and easier to cost out. The more complex the investigation process becomes, the harder it is to anticipate all the procedures that will be needed and their associated costs. Fortunately, by the time the complex diagnostic stage is reached, management is familiar with and comfortable with the diagnostic team; so the uneasiness that comes with wading into uncharted waters can be avoided. If the comfort and trust isn't there, serious consideration should be given to getting another diagnostic team.

Applied evaluation of IAQ is an emerging field. Procedures traditionally used in industrial hygiene are inappropriate for non-industrial applications. Ignoring the wealth of information IHs have, however, would be foolish. Similarly, neither the medical profession, nor the engineers have the breadth of training and experience to go it alone. Ultimately, the team must provide a comprehensive and cost-effective approach to either resolving or preventing building-related problems. An effective team integrates the disciplines of science and engineering along with subtle interactions between occupants, buildings systems, and facilities management. It is equally important that the diagnostics team have a strong, competent, and communicative leader to interpret and interface with building management. This maximizes the probability that the problem will be brought to a timely and cost effective resolution.

Chapter 5

Controlling Indoor Air Problems: How to Keep the Building Working Well

In practice, there is little to distinguish the control of contaminants from their treatment and mitigation. Treating an existing problem through mitigation or remediation is often the same exercise as preventive measures taken to control or avoid it in the first place. And once treated, the same actions become control procedures to prevent a recurrence. Some preventive actions, of course, can be taken that are not available in remedial treatment. Building design is an example. The full recognition of the importance of designing for IAQ acceptability is recognized and more fully elaborated in the ASHRAE Indoor Air Quality Advanced Design Guide (IAQDG) published in 2009.

Unless specifically mentioned, as in the new construction discussion, prevention and control strategies will be considered as one and the same in the following discussion.

TOXICITY AND HAZARD

Managing indoor air quality requires a basic understanding of toxicity and hazard. These terms are frequently used incorrectly in the press. They are not synonymous. Every substance is toxic to humans in some manner, or to some degree, even orange juice. A toxic substance becomes hazardous by intensity of the dose, the length of exposure or the manner in which it is introduced into the body. Few would fear orange juice in a glass, but no one would want it injected into his or her blood stream.

Hazard is actually a measure of risk or the probability that an unwanted event will occur. Hazard is usually measured in severity or magnitude. It is a product of toxicity and exposure (or dose). Exposure is usually defined as *acute* when it is intense for short duration, and

chronic in the case of low level over a long period of time. Control procedures reduce or eliminate the level of toxicity, the dose, or both; thereby,
reducing the hazard or risk.

MEASURING CONTAMINANTS

In the process of controlling a contaminant, if it can't be eliminated, then some gauge of acceptable level is desirable. Building owners
and facility managers, in most instances, simply do not need to know all
the particulars about acceptable levels, measurement techniques, or the
jargon that goes with it. When management gets embroiled in a serious
problem, they soon learn more than they ever wanted to know about
that particular pollutant.

Fortunately, about 80 percent of the problems associated with indoor air quality can be identified with a walk-through inspection using
simple diagnostics and an experienced and educated eye. Extensive
testing is used in approximately 20 percent of the SBS investigations
and the definitive cause is found in most but not all of those. The difficulties associated with investigations, as discussed in the preceding
chapter, help explain why some cases are never resolved completely.
Testing for contaminants clearly meets the test of diminishing returns:
it is progressively more expensive while the possibility of positive results declines.

The difficulties inherent in measuring contaminants, the limited
talent available to analyze findings, and the high cost of measurement
argue for an effective control and prevention program. Since such a
large percentage of the problems can be found in a relatively simple
walk through, control opportunities are relatively easy to identify.
Many only take a little common sense. If investigations and interviews,
for example, reveal three findings: (1) occupants experience and report
the symptoms associated with bioaerosols; (2) the area was recently
flooded; and (3) the carpet got wet, the logical answer is: "replace the
carpet." The alternative, in almost every case, would be to go through
an exceedingly costly and elaborate process of testing for bioaerosols,
only to end up with a recommendation: "replace the carpet" after the
delay of several weeks of increased exposure and reduced productivity.

CONTROLS

To accommodate occupant needs, building owners and facility managers must recognize the multiple purposes of controlling the quality of indoor air. First of all, federal requirements for specific contaminants must be met. Furthermore, satisfying guidelines that do not have the force of law may still be financially and legally prudent as a defense against negligence in potential lawsuits.

The underlying purpose in all control procedures is to be sure the indoor air:

• maintains the quality needed for safety and health;
• satisfies comfort and productivity needs; and
• is as cost-effective and energy effective as possible.

The variability of organizational and occupant needs further complicate control procedures. The "status quo" is never static. As facility managers and operating engineers have long recognized, seasonal variations call for accommodations in air conditioning (heating and cooling) and in lighting. In fact, controls to meet these needs vary from day to day, even hour to hour. Similarly, air quality needs may vary by tasks and the scheduling of those tasks. For example, VDT use may be much greater at the end of the month. Or, copiers may suddenly be used extensively just before the big meeting.

Obviously, operators could be run ragged trying to meet every little variation. The goal is to avoid rigidity in implementing control strategies and to allow sufficient flexibility and resiliency to meet these variables. As more sophisticated automated systems emerge, their sensing/control devices will facilitate more subtle responses to individual needs and conditions.

Control Methods

Methods for controlling fall into only three categories: (1) elimination at the source(s)—source control; (2) dilution with less contaminated air—ventilation; (3) extraction with some type cleanser—filtration. They can be further broken into more precise measures of design, operation, or maintenance to help identify the needed control actions. Many control procedures were first developed in the industrial workplace. While industrial hygiene and safety considerations are beyond the purview of this book, some tactics are mentioned here because of their carry-over

implications to other occupational settings. The following listing provides greater detail on the wide variety of tactics available for prevention and control of contaminants. Considerably more detailed guidance
is available in the ASHRAE IAQDG, especially in regard to availability
and evaluation of low-emission materials and the selection of appropriate filtration and air cleaning equipment.

1. *Elimination*—the complete removal of; (a) the biological agent,
   (b) a toxic substance, (c) a hazardous condition and/or, (d) the
   source.

   Elimination procedures include maintenance actions to remove the breeding grounds for bioaerosols, the removal of friable asbestos, the banning of smoking, or the use of air cleaning
   devices at contaminant sources.

2. *Substitution*—the deliberate purchase or use of less hazardous
   materials; e.g., pesticide selection, the purchase of low emitting
   furnishings or building materials, the selection of latex/water-
   based paints over oil-based paints wherever possible. Review of
   MSDS information will aid in the selection of low out-gassing
   product. Partly driven by the initial "Carpet Dialogue" with
   EPA, the Carpet Institute has developed a Green Label program
   that reports out-gassing data on carpeting products. Driven
   further by the Green Building and other labeling trends, such
   as "GreenGuard," other product manufacturers are providing
   testing data and additional content information that is helpful
   in selecting "indoor environmental friendly" products.

3. *Isolation*—containment, encapsulation, shielding, sealing, timing, and the use of distance are all means of isolating a contaminant or a source from exposure to humans. Distancing may be
   accomplished through location and/or time of use. Examples
   of isolation controls include painting or insect treatment during
   unoccupied hours, asbestos encapsulation, and removing orna-
   mental plants from the facility before spraying with pesticides.

4. *New construction/renovation design*—many design steps can be
   taken to prevent problems from occurring including ventilation
   effectiveness, thermal comfort, lighting, the selection of building

materials and maintenance needs. Filter selection and access, for example, is an often overlooked but critical design consideration.

Procedures to commission a new building, including extensive air purging or "flush-out" and occupancy schedules, should be considered. (See section on bake outs later in this chapter.)

Modifications to buildings to be avoided include changes that restrict original air flow design, such as partitions in an open space facility, or changes that increase heat and contaminant load beyond HVAC capabilities.

5. *Product or process change*—while borrowed from industry, change in process controls have applications in other settings. Component design to reduce emissions might be used, for example, to affect the way a Zamboni machine is exhausted at an ice rink.

6. *Housekeeping and dust suppression*—actions that keep surfaces clean of contaminants, prevent their redispersion, and/or eliminate personal contact entirely are important control measures. Some very common place controls are windbreaks, care in preventing vacuum cleaner leaks, improved vacuum cleaner bag performance to HEPA level for example, and efforts to contain/ isolate dusty sources, such as printers and paper shredders. Walk-in contaminants should also be addressed and some excellent guidance is presented in the ASHRAE IAQDG under the section dealing with Track-off systems at entrances.

7. *Maintenance and work practices*—specifications for the proper work procedures to reduce or control contaminant releases for purposeful reasons, such as pesticides, need to be spelled out and should be part of training procedures. Maintenance practices (many of which cut across other control procedures listed here) are vital, especially in the automated control and HVAC areas.

8. *Replacement*—insulation, carpeting, wall coverings, etc., which when wet can serve as breeding grounds for microorganisms, need to be checked regularly and replaced immediately when damaged.

9. *Education, training, labeling and warning procedures*—some training, labeling and warning procedures are required by law.

Whether required or not, workers and management must be educated as to the nature of hazardous materials and ways to minimize risk in their use. In some instances, educating building occupants, guests or public may be necessary.

Owners may need to force some education on professionals, through specification or professional qualifications. The owner carries the ultimate responsibility to see that architects and engineers use appropriate designs and select safe building materials.

10. *Sanitary procedures and personal protective devices*—the use of hygienic principles to reduce or eliminate hazardous materials from a person may be critical under certain conditions. In a hospital setting, protective procedures are accomplished through clothing changes, showering, chlorination, etc. Protective devices are usually used where other control devices are not technically or economically feasible; e.g., respiratory protective devices when O&M people are working with microbial contamination.

11. *Storage and disposal*—laws prescribe the storage and disposal of some contaminants. Care should always be taken to use and store materials as indicated on the label. When in the slightest doubt as to proper use, storage or disposal, good control practices dictate contacting manufacturers, state and/or federal agencies *first*.

12. *Filtering and air cleaning*—the use of adequate and properly selected filters and purification devices in the air distribution system, with outdoor air and mechanically recirculated air, is an essential control factor. Filters & purification devices appropriate to the need should be used and maintained/replaced on a regularly scheduled basis. Generally, the conventional 1-2 inch disposable throw-away furnace filter does not provide adequate cleanliness protection for either the system or the occupants. In new construction, Standard 62 currently requires a minimum filtration efficiency level of MERV 6 on any system having a wet coil. Recommendations are pending to increase this minimum to MERV 8 because field research by Burroughs has indicated that the MERV 6 filter does not assure distribution system cleanliness. Design guidance for more advanced IAQ attainment is recommending MERV 13 minimums along with more elaborate filter seal and by-pass avoidance.

13. *Ventilation*—through increased outside air or exhaust with controlled make-up air. Dilution is a preferred control when the contaminant/source is unknown, source treatment is too costly, or when the source is not localized. Ventilation control means much more than the amount of outside air brought into a facility. It includes the quality of outside air; the effectiveness with which it reaches occupants; and its efficiency in reducing contaminant levels. The opportunities and limitations of ventilation as a control are treated in greater depth later in the book.

Management may also wish to seek the advice and counsel of medical experts as a control support through medical surveillance or treatment. Medical surveillance may include occupant preplacement screening that will restrict high risk persons or provide medical exclusion; e.g., reassignment by location or task of chemically sensitive individuals. Building related illnesses (BRI) by definition depend on medical verification, which in turn usually suggests the cause and the control measures to be taken.

Overriding all of these control procedures are some administrative and management considerations as well as judgment calls. Personnel rotation or scheduled relief breaks may be necessary to reduce the time of exposure to an essential process. As an example, equipment requiring relatively high power charges, such as computers, put positive and negative charges on tiny airborne dust particles; thus, attracting some while pushing others in the face of the operator. That's why computer and TV screens get dusty even on the lower side of convex surfaces. Unfortunately, the same amount of dust that is forced into the air, computer operators must breathe. These concerns and others related to the use of monitor screens may require task rotation and consideration of female workers who are pregnant. To assure control procedures are planned and executed properly, formal organizational steps to delegate authority and responsibility must be made. These management considerations are discussed in the management chapter.

SPECIAL CONTROL CONSIDERATIONS

The importance of thermal conditions to occupant comfort and the relationship to particular contaminants is well documented. The role of the HVAC system in delivering the required thermal conditions and meeting IAQ ventilation needs is a critical IAQ concern. The fact that the HVAC system is also a source of contaminants is a key consideration. Because of their unique characteristics and importance to IAQ, thermal comfort factors and ventilation are discussed separately in later chapters.

Other control factors of particular concern are filter selection and maintenance, purchasing, and new construction/modifications design. These areas are treated very briefly below from the management perspective. Filtration technology is also discussed in its own chapter.

Filter Selection and Maintenance

The primary purpose of filters is to reduce the contaminants in the airflow. The effectiveness with which they fulfill this purpose is determined by the type of air cleaner used, location in the system and the amount of air which passes through. The more effective the air-cleaning system is in removing contaminants from the already conditioned return air, the more that air can be substituted for outdoor air; thus, reducing the energy costs. The amount of filterable material that can be removed is determined by the volume of flow through the cleaner and its overall effectiveness.

The selection of an air-cleaning system is based on the contaminants to be removed, such as the size of particles, and any specialized needs that must be met. For example, a special air scrubber system is used to remove carbon dioxide (CO_2) in the closed system of a nuclear submarine; however, its high cost and maintenance needs make it impractical for ambient conditions in commercial buildings. In other circumstances, the centrifugal cyclone separator is dependent on the density of the particles, their size and the density of the air. Since very small particles cause little aerodynamic drag in a moving airstream due to their surface area versus mass, cyclone separators only have application in settings where larger particle removal is required, such as dust removal in industrial plants.

Building HVAC systems typically rely on arrestance media filters and electrostatic air cleaners. The media filter is a porous material that strains out particles as air is forced through it. Media filters range from the more crude "duststop" filters still used in some warm-air furnaces to the high efficiency particulate arrestor filter (HEPA). The HEPA filter is highly capable of removing submicron particles, which is why they were called "absolute" filters early in their history.

The effectiveness of media filters changes as particulate matter builds up. If left alone, they gradually increase in collection efficiency. That, however, is not a good reason to leave them unattended. As they clog, the flow rate decreases. Resistance from clogged filters reduces air flow and system efficiency and increases energy usage.

Electrostatic air cleaners (EAC) have demonstrated very high efficiency in removing particles in the range 0.01 micron to 5.0 microns. The EAC acts by charging the particles passing through, causing them to be attracted to oppositely charged plates. The collected material must, however, be regularly removed from the collector plates. EACs operate best when absolutely clean. Build up on plates not only reduces effectiveness, but the units can become odor sources. In small installations, the whole filter is removed for washing. Larger EACs are designed with wash-in-place capability. These types of equipment are complex and require high levels of maintenance; thus, they are now only used in specialized applications, such as kitchen hoods.

Gases and vapors can be filtered using adsorbers or absorbers. Activated carbon, with its large surface area due to voids within its structure, is a good example of this type of system. Adhesion of the gas molecules to the carbon removes the contaminants from the air. This process works fairly well on large molecular weight molecules, such as solvent (VOC) molecules, but does not do an effective job of removing small molecules, such as sulfur dioxide. For control of the latter compound and other more reactive gases, a potassium permanganate treated alumina pellet works better. A full range of sorbents, such as carbon, alumina and zeolite, can be treated with reagents to enhance their abilities to capture and retain specific chemical compounds.

The source of contamination can influence the placement of filters in the system. If the source of contaminants is the conditioned space, the air cleaner can be placed in the recirculating airstream or the mixed airstream. If the main source is outside, then outdoor air filtration is a primary consideration. Unless the staff has particular expertise in this area, it is wise to consult a specialist when filter design or modification decisions are needed.

Filtration and air cleaning can be a truly cost-effective and useful tool in IAQ prevention and mitigation. Yet, it is an under-utilized technology in the IAQ arena as the subject of filtration is complex, highly specific and, unfortunately, there are few generic non-biased sources of information. This is why the subject is covered more thoroughly in a separate chapter to provide the IAQ professional with a valuable information resource. Refer to the filtration chapter for further detail on application, product types, testing methods, and design techniques.

Purchasing

Caulks, sealants and adhesives have various levels of VOC emissions depending on the compound. Building, pipe and duct insulation with ureaformaldehyde insulation or asbestos-containing acoustical, fireproof and thermal insulation can have an adverse effect on IAQ. Paints have highly volatile mixtures of VOCs and have high release rates during the short-term curing process. Many of the problems associated with these products can be avoided through selective purchasing.

The careful selection of construction materials and furnishings can help alleviate indoor air problems at the source; e.g., the level of formaldehyde emission from pressed wood products. Table 5-1 lists materials of particular concern that warrant careful selection procedures. Materials of concern encompass site preparation, envelope construction, mechanical system and interior finishing.

Reference in the specifications to certain federal or industry guidelines can avoid major headaches later on, both figuratively and literally. In purchasing materials directly or through the contractor, standards can be specified so that materials meet federal regulations. Manufacturer's Safety Data Sheets should be required of all vendors as per OSHA mandate. But further, the MSDS data should be scrutinized closely for odorous or irritating components and the documents retained in an orderly reference system in case VOC complaints occur. An additional precaution is to ask that each chemical or material in a composition product be identified, and outgassing rates be provided when known.

New Construction/Renovation Design

All materials pollute, some a little, some a lot, but they all contribute to a deterioration of the air quality. "Engineering out" materials and conditions known to have a particularly adverse effect on indoor air quality during new construction or renovation may prove to be the easiest control measure of all.

Now that it has become evident that a building can be hazardous to one's health, the reduction of the potential hazards become a critical criteria of building design and construction. The ASHRAE IAQDG is

Table 5-1. Building Materials of Particular Concern

SITE PREPARATION AND FOUNDATIONS
> soil treatment insecticides
> foundation waterproofing, especially oil derivatives
> high levels of dirt and dust

ENVELOPE
> wood preservatives
> concrete sealers°
> curing agents
> caulking, sealants, glazing compounds and joint fillers
> insulation, thermal and acoustical
> fire proofing materials

MECHANICAL SYSTEMS
> duct sealants
> mastics

INTERIORS AND FINISHES
> subfloor or underlayment
> floor or carpet adhesives
> carpet backing or pad
> carpet or resilient flooring
> wall coverings, both vinyl and fabric
> adhesives
> paints, stains
> paneling
> partitions
> furnishings
> ceiling tiles
> gypsum dust
> window coverings

Adapted from Indoor Air Quality Update, Arlington, MA

an invaluable aid in designing acceptable IAQ into the building at the beginning because it integrates architecture with function and covers in depth all aspects of integrating IAQ into the design/construct process.

Put It in the Specs

The owner, who is ultimately responsible for the facility, cannot assume an architect or an engineer fully understands the design and materials implications associated with indoor air quality. It is incumbent upon the owner to include sufficient direction in the specifications to obtain a facility that is energy efficient, aesthetically pleasing, productive and hygienically safe. Appropriate IAQ specifications offer a quality control opportunity, that can save millions in remedial measures, and could lessen legal liability.

Facility Siting, Foundations

Siting of a facility should consider potentially negative influences on IAQ; e.g., the quality of the soil, particularly in radon affected areas. The proximity of roadways and vegetation may, depending on the type and placement, serve as a source of allergens and high particulate count.

The nature and location of polluting activities should affect design. For example, exhaust near a localized activity may serve as an effective source control mechanism. If theatre or craft production, car repair work or maintenance operations are going to involve toxic gases or particulates from brazing, welding, cutting or soldering, then direct venting to the outside should be part of the design.

Precautions should be taken in new construction to thwart radon entry through slab plate penetration. Recommendations made by Brennan and Turner several years ago in "Defining Radon" still provide excellent control guidance. They recommended: (1) slabs and basements be poured with as few joints as possible and poured right to the wall; (2) wire reinforcement be used in slabs and walls to help prevent future cracks; (3) seams and perimeters be caulked with polyurethane; (4) dampproofing, sealing and coating the walls be done to slow entry; and (5) sub-slab drains and several inches of #2 stone be placed below the slab during construction to provide sub-slab ventilation potential.

Thermal Comfort Design

In addition to the consideration given to the thermal environment and humidity, air flow patterns need careful HVAC designer attention. Designers historically were concerned almost exclusively about temperature control. For years, humidity controls were ignored except in areas of extreme climate or special conditions. While we've provided "climate controlled" environments for rare books, masterpieces, and special documents for many years, we have only recently begun to recognize the importance of humidity in a comfortable, hygienically safe environment. The role humidity plays in fostering/controlling contaminants is being more fully appreciated.

We have not progressed, however, in making sure that the outside air or the cleaned, recirculated air is actually reaching the occupants. With the advent of sealed windows, occupants became totally dependent on the designer's understanding of air flow. To illustrate one renovation problem, we have only to look at the partitioned "open space" offices and schools of the 1970s where pockets of dead air have been created.

Ventilation Effectiveness Design

A more pervasive problem has been revealed in studies which found that most office buildings had both the supply air outlets and return air inlets located at ceiling level into a ceiling plenum. This placement can cause a short circuiting and poor distribution of air. As the air pours across the ceiling, half of the room—the half where the occupants are—is left with poor ventilation. This is exacerbated by poor louvre design and the part load operating condition of VAV systems.

An owner does not have to understand complex formulas on mixing efficiency to be able to spot grill work at the ceiling level and to see why the room still seems "stuffy" even with increased outside air. Such designs also severely constrain the value of ventilation as a control measure. Ventilation effectiveness, the ratio of outside air reaching occupants compared to total outside air supplied to the space, determines the capability of the supply air to limit and control the concentration of the contaminants. Design considerations are implicit in the discussions of temperature, humidity and ventilation effectiveness found later in the book.

Maintenance Considerations

The importance of providing access for maintaining all equipment cannot be overemphasized. Failure to allow for maintenance access occurs entirely too often. The resulting horror stories are legion. For owners and managers, this should be a critical design concern.

Sources of contaminants and their controls, as discussed elsewhere in this chapter and throughout the book, suggest further design considerations. For example, the material on bioaerosols discusses the problems associated with the growth of microbials in the space, in the air handler, and in the distribution system. This is brought on by the improper maintenance and operation of systems that enables water and dirt to accumulate and promote this condition.

Bake Outs

Bake-outs were frequently suggested early-on as part of the commissioning process to control VOC contaminants in new construction. Elevated temperatures were thought to speed VOC emissions. The bake-out was, therefore, expected to accelerate the off-gassing process prior to occupancy and thus reduce the pollution levels after the building is occupied. A bake-out was controlled by adjusting three parameters; the duration of the bake-out, the indoor air temperature during the bake-out, and the ventilation rate during the process. In 1989, J. Girman and his colleagues reported reduction of selected VOCs as a result of the bake-out process. He also confided in verbal reporting that the research building suffered substantial physical damage to wall coverings, concrete foundation, and millwork. Researchers, headed by C. Bayer, at the Georgia Institute of Technology, later reported that total counts of VOCs were "about the same before and after the bake-out," but they found that numerous specific compounds had shifted to other totally new organic compounds. Bayer theorized that the higher levels of VOCs had transported to sorption sites and from these "sinks" re-emitted sometimes as new compounds. The bake-out process must be performed after the building is totally furnished and ready for occupancy. Yet, there must be no occupancy during the process. This creates tremendous financial burden to the owner who must delay the useful start time of the asset.

A successful alternative to bake out is the tactic of "flush-out." This employs as much mechanically induced outdoor air as possible during the final build-out when contaminant loads are highest. Flushing is performed continuously around the clock for several weeks into the move-in and occupancy. One caution is that the technique is to be used only during mild low humidity seasons or with the refrigeration cycle operative to avoid humidity build-up in the space during off occupancy hours.

Emerging Technology

Finally, owners should keep their eye on emerging technology, particularly in the area of automated sensing controls and ergonomically correct furniture. Also, raised floor and underfloor distribution is adding to the ability of occupants to control discharge of supply air into their space while enhancing the overall ventilation effectiveness of the system.

New, more precise, and dependable sensing controls will allow much more accurate monitoring and response to specific constituents such as carbon dioxide. This will enable better demand control of ventilation levels that can be modified in response of varying occupancies. New heat-pipe technology is enabling more efficiency salvage of heat content of air discharged from the building. Newly developed high performance desiccant wheels based upon improved desiccants will aid in humidity control tactics without the use of CFC refrigerants. High flow and high capacity particulate filtration medias and fabrication techniques are making new high efficiency filters available that combine long life with low energy usage. Similarly, there are new developments in gas phase filtration employing sorbent imbedded fabrics that enable medium efficiency performance with lower cost and pressure drop. Improved air flow louvers are becoming available that can assure good air diffusion and mixing even at reduced velocities typical of part load conditions. Advances in UVGI (Ultraviolet Germicidal Irradiation) light bulbs and equipment have enabled the usage of this mechanism in HVAC equipment to control microbial growth on wetted coiling coil surfaces. Unfortunately, the process has limitations in a moving air stream to affect the air quality because of limited dwell time, according to research performed by RTI (Research Triangle Institute). There is also considerable funding emanating from Homeland Security that will bring about other rapid advances in contaminant monitoring and control materials and devices due to potential terrorism concerns.

Managers can't keep up with the state-of-the-art on all technological fronts. Keeping tabs on the continuing emergence of controls, equipment and software to satisfy IAQ needs, reduce energy usage, and optimize productivity and space utilization would be almost impossible. The obvious benefits, however, warrant a thorough investigation of new products and technologies as construction/renovation project planning begins.

Though not an emerging technology, the process of *commissioning* is emerging as a critical component of successful completion of the construction process. It is even more critical in bringing remediation and renovation projects to successful closure. Described in ASHRAE

Guideline 1, the process involves the aggressive overview of each stage of the construction project by a designated commissioning agent or staff person (not the contractor). From start to finish, the process is designed to assure the conformance of the project to the design resulting in a building that performs according to intent. Along the construction pathway, the process reveals and remedies problems that normally get built-in and concealed only to emerge later as contributors to problems. In commissioning of IAQ projects, both Milam and Woods both reported construction efficiencies to more than offset the cost.

CONTROL EVALUATION AND MONITORING

Control management requires some evaluation of the steps taken. The level of evaluation will vary with the condition(s) which caused the problem. With sick building syndrome (SBS), where the cause may never be identified, the cessation of complaints may serve as sufficient evaluation of treatment procedures. In the case of something as alarming as Legionnaire's disease, medical verification and a proactive program of analysis of water sources for *Legionella* may be warranted.

Following the treatment for SBS or BRI, a critical component of effective control management is monitoring. Monitoring may range from a simple oversight procedure that assures continued adherence to control practices, to intermittent or continuous atmospheric sampling and analysis methods.

Sampling and analysis for hazards may include area, personal, process, or duct sampling to determine characteristics of emission, level of exposure and operational conditions. Sampling and analysis procedures are usually very specialized, conducted by outside consultants and are expensive. It is an evaluation procedure generally reserved for extreme circumstance or research.

Currently, the most widely used monitoring approach to assure continued quality of the indoor air is monitoring carbon dioxide (CO_2) measurements. As discussed previously, CO_2 is frequently used as a surrogate for other contaminants. Except in instances where sources, such as vehicle exhaust, affect CO_2 content, the concentration of CO_2 in the outside air is fairly constant. With information about ambient CO_2 levels outdoors and what is going on indoors, we can learn a lot about the effectiveness of the ventilation system.

If CO_2 sensors are used to monitor concentrations, care should be taken to be sure they are sufficiently reliable and hold calibration. If they drift more than 100 ppm per year, then they will not offer a reliable reading nor accurate control if being used for variable demand controlled ventilation on a permanent basis.

CONTROL BY CONTAMINANT

The effectiveness of specific controls varies by contaminant. Ventilation as a control device, for instance, will not benefit and may aggravate certain contaminant situations. When the pollutant is known, controls should be contaminant-specific. Some control measures are required by law. In such instances, legal reference is made to the applicable law or code in the following discussion of controls by specific contaminants.

Asbestos
Abatement methods are:
1) Operations and maintenance
2) Repair
3) Enclosure
4) Encapsulation
5) Removal

The Asbestos Hazard Emergency Response Act (AHERA) is a federal law that specified acceptable abatement procedures. Some people falsely assumed that AHERA required asbestos removal. The law only required schools to exercise every effort to protect human health and the environment by "the least burdensome method." Financially, the least burdensome method seldom equates with removal. Furthermore, as the Harvard report, *Summary of Symposium on Health Aspects of Exposure to Asbestos in Buildings*, reminds us, "Removal itself is not without risk."

The threat of lawsuits prompted building owners to opt for removal; however, removal has the potential to increase, rather than decrease, indoor air concentrations. Removal and disposal exposes some workers to high concentrations of airborne asbestos. Owners may inherit an even greater liability through removal, for they are responsible by Federal Regulations for the asbestos at the disposal site for 40 years.

Abatement specialists, fully trained with appropriate credentials and experience, should be employed for asbestos detection, encapsulation, removal, disposal and air monitoring procedures. Owners should seek protection by having specialists bonded and insured. State counterparts of EPA and OSHA can help identify appropriately qualified contractors.

When reviewing contractor credentials, try to retain the services of specialists with experience in like facilities. Contractors, who gained their experience in schools, may not appreciate the difficulties to be found in government offices, hospitals, or multi-tenant high rise buildings where the HVAC system can't be shut down or the space totally evacuated during work.

Asbestos removal, when the HVAC is not shut off, presents a whole new set of conditions. Asbestos removal generates high levels of asbestos dust that may be transported in the HVAC system. Preplanning, by a team which includes a mechanical/HVAC engineer with asbestos expertise and an industrial hygienist, will help determine the precautionary steps to take; e.g., work area isolation, negative HEPA air filtration, personnel and area decontamination. It may be necessary to weigh the viability of the continued operation of the HVAC system.

OSHA guidelines for asbestos removal can be found in Title 29, Code of Federal Regulations, 1910.1001 and 1926.58, U.S. Department of Labor, Occupational Safety and Health Administration, 1993. It is management's responsibility to see that these guidelines are followed.

Bioaerosols/Molds

A review of microorganism and their sources suggests ways to control bioaerosols and microbial contaminants. Sources for biological growth include wet insulation, carpet, ceiling tile, wall coverings, furniture and stagnant water in air conditioners, dehumidifiers, humidifiers, cooling towers, drip pans and cooling coils of air handling units. People, pets, plants, insects, soil may carry biological agents into a facility or serve as potential sources. Frequently, bioaerosols settle in the ventilation system itself where viable spores and bacteria can colonize and grow. The inevitable dust inside duct work plus condensate moisture (often resulting from variations in temperature and humidity) can work together to turn these vast surface areas into breeding grounds for mold. Because of numerous noteworthy and well publicized mold cases, mold has become the feeding frenzy of lawyers, investigators, and mitigation firms. Several high profile settlements involving celebrities and millions of dollars has fueled the tort litigation factions. Names like the Ballard Family, Erin Brockovich, and Ed McMann have become poster children of this issue and synonymous with huge multimillion-dollar settlements in mold litigation. Thus, mold has emerged as the building owners' scourge of the 21st century.

To show that this is almost a timeless exercise, the following quotation is a tongue-in-cheek distortion of scripture taken from Leviticus 14:33-59 with credit to Don Gatley, a noted IAQ expert in humidity control.

> *"The Lord also said to Moses and Aaron, when you come into the land of Canaan unto your people, when I might cause mildew to grow in a house in the land of your possession. Then he who owns the house and finds mildew, shall come to the Priest and indicate that he found mildew. The Priest will command the house to be emptied lest all that are in the house be declared unclean. Then he will go and examine at the mildew to see if the mildew is on the walls of the house with greenish or reddish spots, and if it might go deeper than the wall's surface. If these things are true, the priest shall go out and close the house for seven days. On the seventh day, he returns and if it has spread, then they must tear out the stones with the mildew on them, they should throw the stones away and put them at a certain unclean place outside the city. Then the priest shall cause all the inside of the house scraped round about and the people must pour all the plaster they scraped off the walls at a certain unclean place outside the city. Then the owner must take other stones and put them in place. They must cover the walls with new clay and plaster. Now, after new stones and plaster have been put in the house, the priest must go look. If after all of this he finds that the mildew has reoccurred, then the house shall be broken down completely, its stones, plaster, and timbers and they shall carry them forth to the unclean place outside of the city. And then they shall start over."*

Although these biblical references were referring to Hansen's Disease (no relation to Shirley Hansen, the author of the original MIAQ) or otherwise known as the feared and contemptible Leprosy, may seem like extreme and drastic myths, all one has to do is examine closely the mitigation tactics used on some notable Florida public buildings, which were rebuilt from the outside inward and again, inside outward at costs far in excess of initial cost. Furthermore, the cleanup process is following closely the tactics used in asbestos abatement, which increase the costs exponentially. This is partly due to early guidance from the New York City *Guideline on Assessment and Remediation of Fungi*; the EPA *Mold Remediation in Schools and Commercial Buildings*; and the *OSHA Guide to Mold in the Workplace*. But also, the issue is being driven by the ex-asbestos contractors who are waiting in the wings ready to leap on buildings in trouble. This brings a new light to the cost effectiveness of prevention and control rather than waiting for the building to "break" before repairing it.

Most remedies and prevention tactics fall in the area of maintenance and preventive maintenance; e.g., cleaning filters and wet areas in the ventilation system; replacing water-damaged carpets, insulation; maintaining an average relative humidity between 30-60 percent; cleaning/disinfecting drain pans and coils; repairing water sources, etc. General cleanliness of the facility and the mechanical systems will satisfy many of these needs and it is important to note that "elbow grease" does not create or emit VOCs!

One of the most critical control issues for the prevention of microbial growth is the stringent and consistent control of moisture and relative humidity as mold growth is only the symptom of a moisture and/or humidity problem. This may require a reexamination of the outdoor air system, refrigeration capacity, coil selection, and control or operating schemes. Ultimate control may require dedicated humidity control equipment or strategies. The important consideration is to avoid the inappropriate dew point and thermal shock occurrences throughout the facility and the HVAC system that causes condensation where it should not occur—on walls, in gypsum board, in wall cavities, behind wall coverings, in building insulation, on vapor barriers—brought on in many cases by improper pressure relationships. There are few legitimate locations for water inside a building—a drinking fountain, a janitor's closet, a commode bowl, a drain P-trap, and a condensate pan (and the latter is only to be a pathway—not a reservoir). Other sites or sources of moisture or water accumulation for whatever reason will result in "bad news" from a fungal growth standpoint.

While good quality filters will remove microbial contaminants, they will not do the job if the contaminants are generated downstream. Stories are told of "rattling the cage," that is, disturbing the air conveyance system and increasing contaminant levels in occupied space fourfold over outside air levels. Cleaning humidifiers, coil surfaces and drain pans are the most effective remedial actions for microbial contaminants generated downstream of the filter banks. The chapter on HVAC discusses bioaerosol-related maintenance in systems.

Filtration of bioaerosols is relatively easy as microorganisms fall in a particle size range that can be effectively trapped by a variety of filters. Most microbial cells range from 1-20 μ with a few from 0.5-200 μ, all of which can be removed in a good quality filtration system. MERV 14 efficiency filters will remove most microbial particulates in the return or outdoor airstreams.

Biocides are sometimes recommended for cleaning ventilation systems. The American Conference of Governmental Industrial Hygienists Bioaerosol Committee recommends that no biocide be used in an *operating* ventilation system. Decommissioned HVAC systems can be cleaned with biocides provided all biocide is removed before the system is restarted. Risks associated with biocides may be worse than exposure to microorganisms. Also, biocides applied to interior surfaces must be registered with the EPA for application in an air stream. Remember that a chemical that can kill a cockroach, a bacterium, or a mold spore will not be gentle and soothing to human lung tissue.

The need to replace duct lining is seldom required. If it is, however, it is often cheaper and easier to replace the complete duct section rather than just the lining. EPA has published research that affirms that the process of duct cleaning may even worsen the problem of biocontaminants in the occupied space. In addition, the new porous insulation duct liners have provided product modifications that make them more resistant to mold and more resilient to duct cleaning.

It remains that moisture control is the most cost effective prevention tactic when dealing with mold and bacteria. When and if mold occurs, remediation must be swift and thorough—starting with the correction of the water incursion. Delays, whether triggered by budget concerns, decision trees, priorities, or just "round-to-its," will magnify the negative impact on the structure, enhance exposures and related negative health effects of occupants or tenants, magnify cost of repairs and mitigation, and expand time delays to full productivity and usage of the space.

The remaining dilemma at the time of this writing is the issue that there are no clear and definitive mold standards—the basic problem of "how clean is clean." The Occupational Safety and Health Administration has not established a Permissible Exposure

Limit (PEL) for mold. Neither has the American Conference of Industrial Hygienists established a safe level, however, several states have recently enacted laws requiring licensure of mold consultants and remediators. The medical scientists have declared that there is no clear linkage between mold exposure and severe adverse health effects, yet concede that low levels of exposure can induce allergic reactions in susceptible individuals. This dichotomy mandates that the building owner take deliberate care and due diligence in selecting and managing the remediation team and his consulting experts. This is because cleanliness targets are largely "expert" driven and getting in bed with the wrong expert will end up costing thousands of tests, months of productivity loss, and millions of dollars in laboratory and remediation fees to attain "clean." Thus, the targets for "acceptability" must be mutually agreed upon between the experts and parties to the complaint before remediation commences—otherwise it does not ever end because "0" doesn't exist in the normal commercial building setting that is exposed to outdoor air. There is currently an ANSI approved standard for mold remediation from the Institute of Inspection Cleaning and Restoration Certification (IICRC) titled, "IICRC S520: Standard and Reference Guide for Professional Mold Remediation." Other organizations, such as AIHA and NADCA, have published documents that include guidance on mold remediation.

Combustion Products

The best control of internally generated combustion products are efforts to maintain, properly adjust and carefully operate all combustion equipment. Vehicular exhaust from garages, loading docks, etc., constitute another major source of combustion products, that needs to be carefully managed and avoided.

The National Aeronautics and Space Administration (NASA) has published research that claims house plants can serve as living air cleaners for the volatile organic chemicals often found in combustion products. From a practical standpoint, it would require extensive foliage in a commercial building to measurably impact the overall level of pollutants. Plantings can also impose a negative IAQ aspect by providing sites for fungal growth due to aged potting soil and careless overwatering and spillage.

When unusually high levels of combustion contaminants are expected in an area, additional ventilation can and should be used as a temporary measure. Conversely, when the ventilation air is the source of contamination, as in peak rush hour periods, the outdoor air may be shut down according to ASHRAE Standard 62. Although carbon monoxide cannot be filtered with current technology, other odorous and irritating outdoor pollutants resulting from combustion can be cleansed using engineered gas phase filtration systems. When combustion appliances are situated within the occupied zone, special care should be taken to avoid negative pressurization, which can cause back drafting of products of combustion into the breathing zone.

Environmental Tobacco Smoke

Even though most public buildings restrict smoking, this remains a significant indoor pollutant, especially in the hospitality industry such as restaurants and bars. Considering that over 20% of the adult population still smoke, the control of the problem contaminant remains a concern to the building owner or manager of facilities that contain smoking designated spaces. Thus, the proper management and control of ETS should be addressed.

The control and treatment of ETS generally falls into five areas:

(1) *remove the source*—eliminate smoking. Decrees to eliminate smoking are policy decisions with employee relation considerations. However, the smoking curtailment proposed in the 1994 Proposed Ruling by OSHA, although formally withdrawn, prompted many state and local authorities to restrict smoking in public space. Thus, today most public buildings restrict smoking entirely or provide designated smoking areas. More recently, many states and local authorities have limited or eliminated smoking in hospitality areas.

(2) *modify the source*—relocate/separate smokers. Separating smokers will reduce, but not eliminate complaints about ETS from non-smokers.

(3) *dilution ventilation.* Increasing ventilation may prove to be the most desirable option, but it is a very expensive one. Standard 62 requires increased ventilation for designed smoking areas. Ventilation rates to satisfy ETS-related health concerns have not been established. ASHRAE 62 guidelines are designed to reduce

tobacco smoke odor and discomfort or irritation, not necessarily reduce health risks.

(4) *filter the contaminants.* HEPA filters and electrostatic precipitators can remove the respirable particles of ETS smoke. Since both odor and irritation are mainly caused by the gaseous phase of smoke, this does not provide adequate control of the contaminant problem. Granulated filter media, such as activated carbon, or permanganate treated alumina or zeolite can collect much of the volatile component of ETS.

(5) *isolate smokers and exhaust contaminants directly to outside.* Establish a dedicated and designated smoking lounge. To assure smoke isolation, the space must be totally sealed from the public space; must exhaust its return directly to the outside; must maintain negative pressurization compared to the adjacent public space; must increase ventilation supply air to a minimum of 60 cfm per smoker; and may supplement with recirculating air cleaners.

Formaldehyde (HCHO)

Through selective purchasing, materials can be obtained with lower potential formaldehyde off-gassing. Researchers have found up to a 23-fold difference in emission from the same products from different manufacturers due to different resins being used and/or pre-treatment to reduce emission levels. Contact and inquiries with dealers and manufacturers with regard to potential formaldehyde emissions rates prior to purchase is warranted and may result in materials with lower exposure levels.

The potential to significantly reduce the irritation from formaldehyde warrants a pro-active, preventive approach. Laminated products should have all exposed surfaces or unused "plug holes" covered with laminate or sealed.

Like other VOCs, out-gassing of HCHO can be lessened by venting the component, such as carpeting and fixtures in unconditioned warehouse space prior to installation.

Barrier coatings and sealants might be used to reduce formaldehyde emissions. Researchers at Ball State University (Godish) have tested various sealants on particleboard flooring and have found they are effective. Barriers, such as vinyl floor coverings, have reduced

HCHO in residences up to 60 percent even with other HCHO emissions present. Barrier coatings and sealants pose their own IAQ problems and adequate ventilation should be maintained during application and until the strong odor fades. Prior to using sealants, notification to chemically sensitized people is recommended.

Formaldehyde is a difficult compound to filter because it does not sorb well onto the common activated carbons. The permanganate/alumina type pellets control the contaminant very effectively. However, a high concentration and/or a high generation rate will consume the sorbent at high rates which can drive upward the air purification costs due to short life cycles.

Radon

The primary sources of radon in buildings with high concentrations is the pressure driven flow of radon soil gas. This pressure difference may be the result of indoor-outdoor temperature differences, prevailing winds, the mechanical ventilation systems and combustion devices that have a depressurizing effect on the building. Even buildings having positive pressure over-all may experience localized negative pressure on the footprint floor due to stack effect or the location of a primary mechanical room. There may also be foundation penetrations because of sumps, drains, or sewage collection. These are potential conditions to allow entry points for radon gas. Care should be take to assure that these penetrations are sealed and the footprint floor operates with positive pressurization so that ground derived gas is not pulled into the conditioned space.

The most effective means of controlling radon is to prevent it from entering the building. In radon areas, precautions should be taken to thwart radon entry. In existing buildings, sealing all cracks and openings around drains is helpful. Some studies have reported mixed success with this control procedure. Facilities with concrete block walls offer many, many avenues for radon entry. Sealing the block walls and ventilating the cavities have cut radon levels up to 90 percent.

In residences, mechanical ventilation of crawl spaces has proved to be very effective. Using fans to draw soil gas from beneath a slab in sub-slab ventilation has been shown to be 50 to 90 percent effective in reducing radon levels. In situations where a basement is fairly tight, pressurizing the basement can be effective. This relatively simple and inexpensive process has reduced radon concentrations 65-95 percent below radon concentration guidelines. As an alternative, increasing ventilation at the first floor level with exhaust out of the basement, as conducted by Wellford in his Pennsylvania study, proved to be effective.

General non-specific increased ventilation is frequently recommended in the literature as a way to reduce indoor radon concentrations. Repeated studies have found no correlation between radon concentrations and air exchange rates. (See Figure 1-1.) In other words, just increasing air exchange rates in general may not work. Ventilation must be applied in a specific fashion to be effective as a control.

Several states require that those who engage in, or profess to engage in, testing for radon gas or the abatement of radon gas be certified or licensed. The state public health, radiation protection office or environmental agency can provide information as to whether such credentials are required; and, if so, provide lists of those who meet the requirements. The EPA's *Radon Measurement Proficiency Report*, which lists firms and laboratories that have demonstrated their ability to accurately measure radon in homes, may be available from the state. The state agency can usually provide guidance regarding the possibility of radon contamination and effective treatment in a particular geographical area. The state can also offer counsel on the availability of effective detection devices and/or services.

Respirable Particulates

Particles are the easiest contaminants to remove from the air stream. Media filters and electrostatic air cleaners are available for use in building HVAC systems. While more expensive to install and operate, high efficiency extended media filters, high efficiency particulate arrestor (HEPA) filters and electrostatic precipitators can remove respirable particulates from the air stream very efficiently.

Appropriate maintenance with scheduled cleaning and replacement of filters is essential. (For further information, see the chapter devoted to filtration.)

Volatile Organic Compounds (VOCs)

High concentrations of VOCs whether odorous or irritating are best controlled by dilution and, thus, increased ventilation is frequently used to reduce concentrations. Janssen, chairman of the ASHRAE committee that developed the original Standard 62-1989, indicated an effort to control these potentially harmful gases prompted the increase move from 5 cfm/person to 15 and 20 cfm/person.

Since increased ventilation is costly and marginally effective, source control tactics are the more cost effective especially with odor problems. For example, selective purchasing of construction materials, furnishings, maintenance and operational materials can avoid or reduce levels of VOC emissions. Metal shelving, equipment, appliances with a powder coated finish, rather than the conventional painted surface, offer essentially no off-gassing.

Materials, such as custodial cleaning materials, should be stored in well-ventilated places away from occupied areas and definitely not the mechanical rooms or closets that are in the return air pathway.

Time of use can be a key factor. Floor wax, for example, has a very high initial emission factor, which is followed by low-level steady state emissions. EPA research has shown floor wax emissions drop from 10,000 μg/cm^2 to about 500 μg/cm^2 in about an hour, and fall below 10 μg/cm^2 in 10 hours. Purposeful activities such as pest control should be scheduled during non-occupied periods to lessen the risk of occupant exposure to harsh and potentially harmful chemicals.

Direct exhaust or additional ventilation should be used for activities known to have high VOC emissions, such as painting and varnishing. These activities should be conducted away from occupied zones whenever possible just as these activities should be restricted to periods of building operation when occupants are not present.

Because VOCs are usually present in relatively large initial concentrations, gas phase filtration is not normally a cost effective control mechanism. However, when used in conjunction with outdoor air purging techniques, excellent results has been demonstrated that are cost effective. Milam reported to an AIA Healthy Building Symposium dramatic reductions of TVOCs when employing gas phase filtration in combination with continuous flushing through the first six weeks of occupancy. The continuous 24 hr. flushing process reduced peak concentrations from over 12,000 micrograms per cubic meter, to an equilibrium target of less than 500μg. The facility has been monitored for over 10 years and low stable VOC concentrations have been maintained. (See Figure 4-5.)

Table 5-2 offers a brief summary of primary control techniques for the major contaminants.

Table 5-2. Summary of Control Techniques by Contaminants/Sources

CONTAMINANT	SOURCES	CONTROL TECHNIQUES
Asbestos	Furnace, pipe, wall ceiling insulation fireproofing, acoustical and floor tiles	1. Enclose; shield 2. Encapsulate; seal 3. Remove 4. Label ACM 5. Use precautions against breathing when disturbed
Bioaerosols/Mold	Wet insulation, carpet, ceiling tile, wall coverings, furniture, air conditioners, dehumidifiers, cooling towers, drip pans and cooling coils. People, pets, plants, insects, and soil. Construction components—wood and paper.	Use effective filters to avoid build up of nutrients. Check and clean any areas with standing water. Be sure condensate pans drain and are clean. Treat with algicides. Maintain humidifiers and dehumidifiers. Check & clean duct linings. Keep surfaces clean and dry. Remove visible mold.
Carbon Monoxide (CO)	Vehicle exhaust, esp. attached garages; unvented kerosene heaters and gas appliances; tobacco smoke; malfunctioning furnaces; contaminated outdoor air.	Check and repair furnaces, flues, heat exchangers, etc. for leaks. Use only vented combustion appliances. Be sure exhaust from garage does not enter air intake. Avoid negative pressurization. Shut down outdoor air during bad peak periods.
Combustion Products (NO$_x$ and CO)	Incomplete combustion process	Use vented appliances and heaters. Avoid air from loading docks and garages entering air intake. Check HVAC for leaks regularly, repair promptly. Filter particulates. Selective gaseous filtration. Avoid boiler exhaust.

Table 5-2. Summary of Control Techniques by Contaminants/Sources (Continued)

CONTAMINANT	SOURCES	CONTROL TECHNIQUES
Environmental Tobacco Smoke (ETS)	Passive smoking; sidestream and mainstream smoke	Eliminate smoking; confine smokers to designated areas; or isolate smokers with direct outside exhaust. Increase ventilation. Filter contaminants.
Formaldehyde (HCHO)	Building products: i.e., paneling,, materials with lower particleboard, plywood urea-formaldehyde insulation as well as fabrics and furnishings.	Selective purchasing of materials with lower formaldehyde emissions. Barrier coatings and sealants. New construction commissioning. Ventilation. Specific gaseous filtration. Selective exhaust.
Radon	Soil around basements and slab on grade	Ventilate crawl spaces. Ventilate sub-slab. Seal cracks, holes around drain pipes. Positive pressure in tight basements. NOTE—Increased ventilation does not necessarily reduce radon levels.
Volatile Organic Compounds (VOCs)	Solvents in adhesives, cleaning agents, paints, fabrics, tobacco smoke, linoleum, pesticides, gasoline, photocopying materials, refrigerants, building material	Avoid use of solvents and pesticides indoors. If done, employ time of use isolation. Localized exhaust near source when feasible. Selective purchasing. Rigid MSDS control. Selective gaseous filtration. Increased ventilation.

Chapter 6

THE THERMAL ENVIRONMENT: THE INTERNAL HABITAT

Historically, the primary purpose of a heating, ventilating and air-conditioning (HVAC) system has been to provide the human organism a comfortable internal environment. When Willis Carrier invented and patented his "Apparatus for Treating Air" in 1906, his intent was to control the indoor environmental climate with part of that objective being the control of relative humidity as his first system was to control Rh in a printing facility in 1902. Exactly what those environmental conditions should be takes on a little different perspective if we first consider the human organism as described so delightfully in engineering terms.*

Man is a complete, self-contained, totally enclosed power plant, available in a variety of sizes and reproducible in quantity. He is relatively long-lived, has major components in duplicate, and science is rapidly making strides toward solving the spare parts problem. He is waterproof, amphibious, operates on a wide variety of fuels; enjoys thermostatically controlled temperature, circulating fluid heat, range finders, sound and sight recording, audio and visual communication, and is equipped with automatic controls called the brain.

What a marvelous machine! The sophisticated complexity of this machine seems to stress how very sensitive it must be to abnormal or hostile conditions. The truth of the matter is that the human body is relatively resistant and able to accommodate environmental conditions that would be totally disastrous for actual machines.

(*Editor's Note: In the original edition, Shirley Hansen graciously quoted the above section from my inaugural address when I became president of ASHRAE in 1987. She asked that I leave the simile in place to set an appropriate stage for this chapter. H.E. Barney Burroughs)

So what precisely does this machine need to be comfortable and efficient... or, healthy?

COMFORT AND HEALTH

A comfortable internal environment for the human organism has traditionally been defined by temperature, humidity and "fresh" air. Health concerns stemming from contaminants in the indoor air are a relatively recent phenomena. We have assumed over time that the environment is healthy if the body is comfortable. And, conversely, in a healthy environment the body will naturally be comfortable. Even Willis Carrier in the first billboard advertising his indoor climate controller in the 1930's claimed the primary benefit of air conditioning would be "improved health."

As IAQ concerns have come under closer scrutiny, we have had to revise our thinking. Health and comfort are not synonymous. An indoor environment can be relatively healthy and still not be comfortable. Our human "machine" may accommodate a contaminated environment where other machines would fail, although it may be uncomfortable. Conversely, an indoor environment can be comfortable and still not be healthy. Occupants may be exposed to health hazards, such as radon or asbestos, and still feel perfectly comfortable. The earlier accepted relative humidity range, 20-80 percent, was based primarily on comfort concerns; the narrower range today, 30-60 percent, incorporates health considerations.

Comfort, or the lack of it, is affected by several personal and environmental factors. Measurable comfort parameters include:

temperature and localized thermal discomfort;
relative humidity; and
air flow; i.e., ventilation.

When temperature and humidity exceed accepted comfort parameters, they can, by themselves, negatively impact air quality and be detrimental to health. We now know that the comfort factors of temperature and humidity can also interact or influence contaminants that may affect health, affirming the multiple nature of the indoor environment i.e. IEQ (Indoor Environmental Quality).

TEMPERATURE

Trying to provide a comfortable temperature for all building occupants prompts a slight paraphrasing of Lincoln's admonition ...but you can't please all the people all of the time. What constitutes a comfortable temperature for one individual may not be acceptable for another person in the proximate area. An individual's ambient temperature needs may vary with age, physiology, activity, clothing, air movement, humidity, uniformity of temperature, contribution of solar gain, number of people in an area and occupant preference.

Moreover, a comfortable temperature may also be a matter of perception. Facility maintenance directors regularly tell the story of removing the thermostat lock box cover in a room where the occupant perennially complained. As soon as the occupant could adjust the thermostat, so the story goes, the person quit complaining. What the occupant didn't know, of course, was that the thermostat had been disconnected. The psychological implications of the human organism's inability to control its own environment could also be a factor in such a situation.

Recommended temperatures are usually given in ranges for the heating and the cooling seasons. In addition, a particular portion of a building; e.g., gymnasium, or an occupant activity, may be specified. The most common gauge of comfort is the ambient temperature. Recommended temperatures typically range from 68°F to 74°F for the winter and from 73°F to 78°F during the summer with the difference being responsive to both variations in clothing and applicable relative humidity. A precise temperature, in and of itself, does not contribute to, or detract from, indoor air quality as long as the occupants are relatively comfortable.

The ASHRAE Comfort Standard 55-2010, *Thermal Environmental Conditions for Human Occupancy*, recommends the use of operative temperatures. The standard considers the ambient air, the mean radiant temperature, clothing and the activity level in calculating desirable operative temperatures. Figure 6-1 displays the ranges of operative temperature and humidity for persons clothed in typical summer and winter clothing at light, mainly sedentary, activity.

Ambient temperatures are, of course, measured with a thermometer. Operative temperatures can be measured with an indoor climate analyzer or a thermal comfort meter. Figure 6-1, however, can be used to determine the desired temperature and relative humidity in an office.

For example, if the ambient temperature is known, the appropriate relative humidity can be ascertained by reading the X axis and following the appropriate line up to the shaded comfort zone and reading across.

The most common complaint related to the thermal environment stems from localized thermal discomfort, where one part of the body

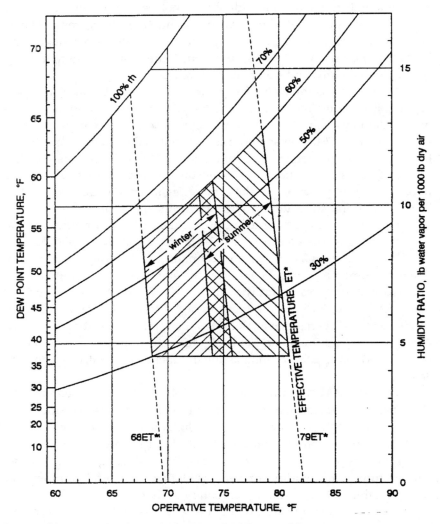

Reprinted by permission from ASHRAE Standard 55, copyright 1992, American Society of Heating, Refrigerating and Air-Conditioning Engineers, Atlanta, Georgia.

Figure 6-1. Acceptable Ranges of Operative Temperatures

is too hot or too cold. Typical occupant complaints are: "My back is too cold," or "My feet are too cold!" Localized thermal discomfort may be the result of air currents or drafts, radiant temperature asymmetry, ground temperatures that are too low or two high, or situations where the vertical temperature difference between head and feet is too great.

There is increasing evidence that productivity and some health considerations are best met in the lower part of the accepted temperature ranges. According to recent studies, higher temperatures can affect mental acuity and are related to the appearance of some SBS symptoms. A reduction in mental work capacity has been observed when temperatures exceeded 24°C (75°F). Significant relationships between room temperatures above 22°C (72°F) and the existence of SBS symptoms have been found in several studies.

Inadequate and/or non-uniform heating/cooling can contribute to indoor air pollution and occupant sensitivity to contaminants.

Higher temperatures and humidity usually accelerates the off-gassing of volatile organic compounds, such as formaldehyde, from building materials and furnishings.

RELATIVE HUMIDITY

Relative humidity (Rh) is the amount of water vapor in the air compared to what it can hold at a given temperature. The level of water vapor in the air affects the body's response to temperature. The Rh level, therefore, affects the temperature ranges found to be acceptable to occupants because it affects the ability of the body to regulate its own temperature through evaporation through the skin. The primary research in this area was conducted by Willis Carrier and published in 1900 while he was an engineering student at Cornell U. It is this relationship that primarily defines ASHRAE's operative temperature.

HVAC designers have historically focused on temperature and humidity. But temperature control alone does not adequately treat all the physiological aspects required by occupants. Neither does ventilation.

Efforts to conserve energy since the mid-1970s have sometimes aggravated Rh problems. Chilled water systems are designed to reduce the humidity from the incoming warm air. As chilled water temperatures were raised to reduce the energy consumption, the amount of humidity in the supply air also increased. This higher humidity level made the air seem warmer to the occupants; and above 70 percent, more uncomfortable. In addition, the likelihood of bioaerosols increased because of the potential of random condensate deposition within components of the structure brought on by thermal bridging.

Research has clearly documented the problems associated with air that is too dry or too moist. The relationship of Rh to the prevalence of mites, fungi, viruses is clear. Airborne viruses, for example, survive best at lower, and higher, humidity levels. Influenza virus survives much better at lower humidities, poliomyelitis virus at higher humidities. The Rh level that is the least hospitable range to contaminants is at about 50 percent.

Morbidity rates for colds have a very seasonal pattern. The incidence of colds increase through the fall, peak in winter and then decline until May, where the rate tends to level off for the summer. Over thirty years ago Lubart, writing in the *New York State Journal of Medicine*, described the relationship of Rh and colds as follows:

> A dry nose and throat caused by artificial heating creates an indoor climate favorable to the cold-inciting agent. Dry nasal mucous is an excellent culture for the infective agent. Adequate moisture is necessary also for the proper function of the cilia, pharynx, larynx, trachea, and lungs. During the heating season the best method of prevention of colds is maintenance of a proper balance of humidity by means of such devices as mechanical humidifier. In summer, regulation of the humidity in air-conditioning is necessary to prevent summer colds.

Before 1960, Andrews found that 40-50 percent Rh with temperatures of 68°F to 70°F produces the most healthy conditions for living and working areas and for recovery from diseases of the respiratory system. He observed that the 40-50 percent Rh "reduces the incidence of respiratory infection and speeds recovery from the common cold." Studies throughout the 1960s and 1970s confirmed the value of proper humidity in the prevention, amelioration and relief of respiratory tract infections.

Despite years of amassing such data, HVAC designers still frequently fail to address humidification needs adequately. If humidity prevents the drying and cracking of wood, leather, paper, etc., why has it taken so long to recognize that the human organism needs humidity, too?

Unfortunately, owners still view humidification/dehumidification as dispensable; therefore, when construction costs come in over budget, Rh control equipment is among the first to go.

With the growing focus on indoor air quality, relative humidity is getting more attention and productivity/absenteeism considerations are apt to expand that attention. Productivity and performance has been shown to correlate with 30-60 percent Rh. This is hardly surprising. Knowing that Rh levels have been tied to the morbidity of colds is reason enough to suggest a relationship to productivity. Find a person with a miserable cold, who feels he or she is functioning at par or better, and you'll be looking at an atypical person. Studies by Ritzel, Sale, Gelperin and Green have all shown a statistically significant reduction in respiratory infections *and absenteeism* among occupants of buildings in which humidity is controlled adequately.

Humidity Control

Relative humidity can be maintained by providing humidification when Rh is too low and dehumidification when water vapor in the air is too high.

Humidification

While air that is too dry presents a range of health problems, humidifier reservoirs that are inappropriate or poorly maintained can create even greater problems. Wherever possible, the humidification process should be limited to direct generated steam injection. However, the steam should be created from heat exchangers using potable or deionized water rather than raw power plant steam that contains potentially harmful amines and biocides.

Assuring proper moisture content in the air can be a money-making proposition, a methodology for calculating the economic benefits of humidifying an office building is offered by Berlin and illustrated below in this discussion.

In general, comfort systems are designed to accommodate the human "machine;" not special machines, such as data processing equipment. A room full of mainframe computers and control equipment may need a system that will respond with sufficient speed to maintain nearly constant temperature and humidity. In instances where building comfort systems are not adequate, computer-based precision air-conditioners that have a sensible heat ratio of 95-98 percent in contrast to the typical

65-70 percent may be needed. Computers throw off more heat and less moisture than people. Precision air-conditions that devote 95-98 percent to cooling and 3-5 percent to dehumidification better meet this specialized need.

Dehumidification

The removal of water vapor from the air is done by mechanical refrigeration or through a desiccant-based system. Mechanical refrigeration has a lower first cost and is more practical down to a certain level. In some instances, it may become necessary to dehumidify at a point where the refrigeration cooling surface would have to be below

---

CALCULATION PROCEDURES:
HUMIDIFICATION ECONOMIC BENEFITS

EXAMPLE: A 25,000 sq. ft. office building has an indoor temperature of 72°F. An electric humidifier to achieve 40% Rh would require 34 kW to produce 103# water/hr. It would need to have a seasonal operating rate of 2,000 hrs. Electricity is $0.10/kWh without demand. Ventilation is 15 cfm/person; 225,000 CFH/O/A. 250 employees have an average salary of $30,000.

PROBLEM: If productivity losses due to a lack of humidity (comfort, respiratory problems, absenteeism, etc.) are 1 percent, what would be the net profit/loss of supplying humidity?

CALCULATIONS:
Grains of Moisture to Add: 3.20/cu. ft.
        (Indoor: 72°F and 40% Rh; 8.59 × 40% = 3.436)
        (Outdoor: 0°F, 50% Rh; .48 × 50% =.24)

Humidification load:

$$\frac{3.2 \text{GRS} \times 225,000 \text{ CFH}}{7,000 \text{ GRS/lb.}} = 102.86 \text{ lbs.hr}$$

Operating cost: 34 kW × 2,000 hrs. × $.10/kWh = $6,800.00
Personnel costs: Avg salary $30,000 × 250 = $7,500,000
        1 % productivity loss = $75,000.00
ANNUAL SAVINGS: $75,000 - $6,800 = $68,200.00

---

freezing to obtain the desired results. When the cooling surface drops below freezing, frost build up interferes with the efficiency of the cooling unit, the unit becomes more costly to run, and the flow of air may be impeded. Under such conditions, a desiccant-based system is warranted. As desiccant equipment costs have come down due to equipment efficiencies and desiccant improvements, this type of control is becoming more cost effective. It is particularly applicable when there is inherent waste heat or economical fuel for regeneration cycles. Bayer *et al.* has reported monitoring successful application of desiccant cooling to schools in the Southeast that were experiencing poor humidity control.

HUMIDITY AND ASHRAE 62

Ironically, compliance with the ASHRAE 62, *Ventilation for Acceptable Indoor Air Quality,* can have a deleterious affect on the indoor air with regard to humidity control in the hot and humid climate. To the extent that ventilation requirements mandate the amount of outside air brought into a facility, it can have a detrimental affect on Rh. In most climates, it will mean more humidity in the summer and drier air in the winter. Since most buildings are already burdened with a design that gave little consideration to controlling humidity, the outdoor ventilation requirements intensify the problem.

The resulting severe decrease in indoor Rh in most buildings during the winter can provoke more colds, respiratory and physiological problems associated with air that is too dry.

In addition to the problems directly associated with respiratory problems and indirectly with bioaerosol growth, water vapor content in the air can significantly affect the release rate of many indoor contaminants and their concentrations in the air. Dry air fosters the breaking of fibers in carpets and other fabrics. Particles are suspended and recirculated for longer periods of time in dry air; thus, increasing the likelihood of inhalation. Further, space that is very dry will promote elevated levels of static charge on components that will affect electronics and cause random deposition these particles on charged surfaces. Higher Rh levels increase off-gassing of VOCs, like formaldehyde or at least heightens the perception of occupants to the contaminants.

VENTILATION

Ventilation is central to indoor air quality. A certain rate of air exchange is essential to a healthy indoor environment. With the advent of the energy efficient, or "tight," building, the drop in infiltration has made mechanical ventilation a more critical factor.

Ventilation as a Control

Ventilation is the most frequently cited IAQ problem and the most frequently prescribed control. Indoor air quality problems and their control are flip sides of the same coin. IAQ corrective measures respond to problems *within* ventilation systems and also *use* ventilation to dilute contaminants.

Ventilation is an effective control procedure, but it is not a panacea. Ventilation as a contaminant control measure is a second choice; and a poor second choice. The control measure of choice is always elimination of the contaminant at the source. Source control through elimination or substitution removes the contaminant threat completely. Dilution by ventilation only "waters" it down.

If a hazardous substance were dripping from a faucet in an occupied area, few would accept hosing it down each morning as the preferred treatment. The hazardous substance would still be there, just diluted. For many contaminants, the long term health effects of chronic low level exposure is not known. Diluting by ventilation only masks the problem and leaves the potential for deleterious effects that could occur through chronic low level exposure. Settling on increased outside air and that alone must be avoided unless the contaminant is unknown and/or the source cannot be determined. Masking potentially dangerous materials that may have long term hazardous effects leaves us far short of what can be done to improve the quality of indoor air.

Much has been made of the National Institute for Occupational Safety and Health's (NIOSH) finding, "In 52% of our investigations, the building ventilation has been inadequate." This frequently quoted statistic has been taken as a mandate to turn up the fan. In this context, inadequate ventilation does not equate to inadequate outdoor air. It is important to go beyond the statistic and look at the "ventilation problems" NIOSH has encountered, which they describe as:

...not enough fresh air supplied to the office space; poor air distribution and mixing which causes stratification, draftiness, and pressure differences between office spaces; temperature and humidity extremes or fluctuations (sometimes caused by poor air distribution or faulty thermostats); and filtration problems caused by improper or no maintenance to the building ventilation system. In many cases, these ventilation problems are created or enhanced by certain energy conservation measures applied in the operation of the building ventilation. These include reducing or eliminating fresh outdoor air; reducing infiltration and exfiltration; lowering thermostats or economizer cycles in winter, raising them in summer; eliminating humidification or dehumidification systems; and early afternoon shut-down and late morning start-up of the ventilation system.

Given this detailed scrutiny, one concludes that inadequate outside air is only one of a long list of ventilation problems. *Dilution is not the 52 percent solution.* Ventilation, problems and solutions, is more than outside air.

NIOSH has pointed out that the 52 percent figure is based on soft data. To the extent, however, that it represents primary problems in the investigated buildings, the NIOSH findings also impart another critical piece of information that is typically overlooked: *48 percent of the problems found by NIOSH will NOT be solved by ventilation changes.* NIOSH has the most extensive experience in indoor air problem investigations and a highly regarded protocol. They have determined that nearly half of the problems they have investigated are not related to ventilation. If the NIOSH data and problems identified by other investigation teams are considered collectively, it seems safe to surmise that a great many of our indoor air problems cannot be satisfied solely by increasing outdoor air intake.

Dilution through increased ventilation certainly offers a means of mitigating some contaminants. Increased air changes per hour may be beneficial when the contaminant or its source is unknown or can't be determined; or, it may serve as an intermediate step until action can be taken. Unique applications of ventilation; e.g. localized source control or sub-slab ventilation to control radon, are valuable control mechanisms.

VENTILATION LIMITATIONS

Ventilation has its place. Dilution is a viable and valuable means of controlling pollution levels. But quantity does not equate to quality. Before we turn up the fan and run up the utility bill, we need to keep in mind that specifying the *quantity* of outside air is only one of many alternatives available and ignores many important factors, including:

(1)   The quality of the outside air may be "bad";

(2)   More air may not help if the ventilation effectiveness is poor;

(3)   There is no single ventilation rate that will assure adequate indoor air quality—other factors frequently have a controlling effect;

(4)   Increased ventilation can be a costly remedy. Costs to move and condition the air represent a significant portion of the utility bill. Increasing ventilation adds to the costs of heating/cooling. To the extent that increased ventilation requires the air handler motor to move more air, the power required to move that additional air is the *cube* of the flow rate;

(5)   Increased ventilation can bring with it serious humidification problems, particularly in the winter months, that may more than offset any gains achieved through dilution; and

6)   The ventilation system may be the source of the contaminant burden.

Ventilation Standards and Non-Occupant Related Pollutants

Ventilation requirements in cfm/occupant is historically based on body odor and smoking. Cfm/occupant does not address contaminant concentrations that are non-occupant related, such as formaldehyde or radon. As with many contaminants, it is the source strength that frequently determines the levels found indoors. For example, variation of radon concentrations and their entry rate has been shown to be much greater than changes in the ventilation rate. Ventilation rate is not the most important variable in radon concentrations in houses. Source strength determines levels found in the building.

Ventilation as a control for VOCs is still unclear. Walkinshaw *et al.* investigated VOC concentrations and ventilation rates in eight Canadian settings; an office, office/lab, office/library, two schools, two hospitals and a residence. Even with ventilation rates exceeding 15 cfm/occupant in all but the schools, the VOC concentrations met or exceeded levels associated with mucous membrane irritations and impaired ability to concentrate.

Hodgson *et al.* measuring VOCs in an office building and a school found that the apparent specific source strength for VOCs approximately doubled with a six-fold increase in ventilation rate. It was not determined what caused the increase in source strength.

Using Ventilation Effectively

The list of ventilation problems found by NIOSH also serves to identify some ventilation solutions that do not require increased air intake. Temperature and humidity fluctuations and extremes as well as filtration problems should be on every ventilation checklist. Common errors in ventilation that frequently impact the quality of indoor air are:

(1) Capture, or "backdraft," hoods located too far from the source of contaminant generation.

(2) Exhaust stacks placed in close proximity to outside air intakes.

(3) The build up of dust on internal surfaces in exhaust systems, which causes increased resistance and reduced ventilation.

(4) Existing exhaust hoods not used effectively by employees.

(5) General exhaust ventilation too often relied on to control employee exposure to airborne contaminants.

(6) The use of flexible ducts in place of rigid ducts, requiring additional energy to move the necessary air volume.

(7) Centrifugal fans operating backwards, which still exhaust air at lower volumes.

(8) Belt driven fans with missing or slipping belts, missing or damaged ducts, which reduce ventilation.

(9) Inadequate make-up air, out of balance with the exhaust system, causing the infiltration of unconditioned and unfiltered air through the envelope and into the conditioned space without treatment.

(10) Addition of hoods and ducts to pre-existing systems without adjustments for increased resistance losses.

(11) Placing walls or partitions in areas originally designed as "open space" without in-wall fans or grill work to help avoid dead air pockets.

(12) Adding copiers, computers, equipment or otherwise changing the functional use of the space without modifying the ventilation to accommodate the changes.

This list, as well as earlier comments, emphasize that design, maintenance and common sense are critical ingredients in operating a ventilation system and in using ventilation effectively as a control measure. Two aspects of ventilation that are often addressed in the IAQ literature warrant further discussion. These two concerns are the quality of the outside air brought into the facility and the effectiveness of the system in delivering the air to the occupants.

"Bad" Air

The quality of outdoor air surrounding a building is affected by external factors; e.g., garbage dumpsters, traffic, loading docks, etc. Designers need to evaluate more carefully the location of intake grills to minimize the outdoor pollution entering the HVAC system and to avoid the building's own exhaust reentering the building.

It is essential that engineers and architects recognize the effects of pressure gradients and air movement around buildings to project airflow particularly in city centers. Careful consideration to the way these factors influence air flow around a building can minimize the entry of contaminated air into occupied building spaces. According to work by Cummins at Florida Solar, as little as 2.5 pa (.01 IWG) negative pressure differential will induce outdoor air into the building. Dependent upon the dewpoint differences between the outdoor and indoor air conditions, uncontrolled condensate can occur within the wall cavity leading to mold growth.

Bahnfleth and Govan reported on the sad circumstances related to a remodeling job when reentry was not adequately considered. A remodeled ground floor of an eight-story building was across the alley from a single-story office attached to a ten-story tower. A mushroom-type exhaust fan was installed in the exterior wall of the remodeled space to serve a steak and hamburger grill. Grease fumes, smoke and odors were exhausted into the alley at the ceiling of the first floor and taken inside through the air intake of the ground floor office.

As concrete evidence of the old adage that a little can go a long way, Bahnfleth and Govan observed that the occasional "whiffs" of cooking steak, which can whet one's appetite, has a negative impact when it becomes a continuous heavy dose of hydrocarbon fumes and odors. Increased absenteeism in one year alone accounted for a $32,000 loss. Lost productivity of those who remained on the job could not be measured but lost effective time was evident.

A centrifugal fan and a short length of duct work were installed to get the exhaust fumes up and away from street level, solving the problem.

Ventilation Effectiveness

The flow of air within the facility and how well that air actually reaches occupants is a concern that increasingly haunts buildings owners and managers. Not only does the absence of ventilation effectiveness cause many SBS complaints, but it frustrates efforts to use ventilation as a control measure.

Most office HVAC systems are designed with the input and output grills near the ceiling. Such an arrangement can cause much of the air to bypass the occupants as shown in the bottom illustration of Figure 6-2. This is particularly true with poorly designed outlet louvers. Also, the part load conditions of VAV systems can drop velocities dramatically allowing air to dump without mixing effectively. This "short circuiting" creates air stagnation at occupant level and fosters complaints of stuffiness and other SBS symptoms. Increasing air intake to the ASHRAE 62 recommended level may create a nice breeze across the ceiling level (at considerable cost), but it may not significantly reduce contaminant levels around the occupants. One problem associated with ventilation as a control measure is the lack of ventilation effectiveness.

VENTILATION AND ENERGY EFFICIENCY

So much attention as been given to reductions in ventilation to cut energy consumption that ventilation and energy efficiency are generally regarded as adversaries. Studies coming out of Europe indicate that methods used to improve the quality of the inside climate also lead to the more efficient use of energy. For instance, Luoma-Juntunen of Finland have found that up to 20 percent of the energy used for ventilation can be saved by the more even distribution of fresh air between the rooms.

Fleming has listed a number of ways that ventilation can be used more efficiently in residences. His suggestions, which have application to other facilities, include:

spot ventilation, using exhaust fans appropriate to the pollution source; use only while pollutants are being emitted:

natural ventilation, opening the windows in mild weather;

fans, which can cool and ventilate less expensively than air-conditioners; and

heat exchangers, an air-to-air heat exchanger can transfer heat from the warm outgoing (inside) air so that 50 to 80 percent of the energy normally lost in the exhaust air is recovered.

ASHRAE predicts a 40 to 50 percent total building energy reduction is possible through the use of heat exchangers.

Fleming's recommendation to "open the windows" in homes may have limited carry over value to other facilities. In an era when "natural" is always assumed to be better and the inclination to throw open the windows is strong, the conclusions drawn by Feustel *et al.* take on added import. The researchers concluded, "Our first set of conclusions from this comparison of ventilation strategies is based on the total airflow and indoor air quality resulting from each strategy. We found that all the mechanical ventilation strategies examined provided more uniform ventilation rates than natural ventilation and, thus, lower total airflow and potentially better indoor air quality."

Weighing natural and mechanical ventilation, Hedges *et al.* found

88 percent of those with air-conditioned offices reported too little ventilation compared with 60 percent of those in naturally ventilated offices. Since the findings were based on interviews, not measurements, it is not clear if occupant concerns were real or perceived. The Danish Town Hall Study reported, "The difference between mechanically and naturally ventilated buildings was not significant for this study." Thus, the question, "Is natural ventilation better?" remains open at this time.

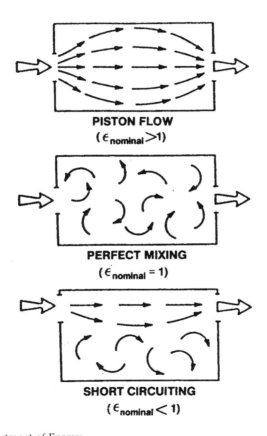

PISTON FLOW
($\epsilon_{nominal} > 1$)

PERFECT MIXING
($\epsilon_{nominal} = 1$)

SHORT CIRCUITING
($\epsilon_{nominal} < 1$)

Source: U.S. Department of Energy

Figure 6-2. Representative Examples of Ventilation Air Movement

The discrepancy in findings may be attributable to the quality of the outside air. Natural ventilation does not offer an opportunity to filter the incoming air. Furthermore, operable windows means occupants have total control of the amount of outdoor air entering a facility. There is no quality control of indoor air with natural ventilation. These uncontrolled openings also disturb the air balance and pressurization relationships which are necessary for good quality indoor environments.

Chapter 7

HVAC: The Heart of IAQ

Since the heating, ventilating and air-conditioning (HVAC) system serves as the lungs of a facility, occupants depend on the HVAC system for comfort, ventilation, temperature, odor and humidity control. The system's effectiveness affects productivity, performance and, most importantly, health.

Unless the HVAC system is well designed and maintained, outdoor contaminants can travel into a building and jeopardize the quality of indoor air. Similarly, the system must filter, dilute and exhaust these pollutants or contaminant levels will increase. The HVAC system is a controller of indoor air quality; and, when not properly maintained, a polluter as well. Poorly maintained HVAC systems are a primary source of pollutants, such as biological growth.

Various studies have established that the HVAC system is responsible for 50 to 60 percent of building generated IAQ problems. The HVAC system is capable of resolving up to 80 percent of the indoor air problems. In other words, if a building is "sick," chances are the HVAC system is at fault, or at least a contributor to the problem and can often remedy the problem. Yet, the design, operation and maintenance of a HVAC system, therefore, is at the heart of a quality indoor air program. As such, it is deserving of special attention.

HVAC DESIGN

A well-designed HVAC system is the mainstay of a healthy building. Poor design is repeatedly cited as a major IAQ problem source. Design related problems are second only to operations and maintenance in contributing to indoor air pollution. More specifically, the authors' and other investigators' experience has found the most design difficulties in:

(1)   ventilation and distribution;
(2)   inadequate filtration;
(3)   pressure relationships; and
(4)   maintenance accessibility.

The design of many HVAC systems in use today originated in the temperature control era. In fact, system designers still tend to be concerned with temperature to the exclusion of humidity and indoor air quality. It is as though design teams have forgotten the psychometrics chart and the role of humidity in comfort. Owners, facing budget constraints, sometimes force this narrow focus on designers.

During the past 20 years, HVAC design criteria have focused increasingly on energy efficiency and conservation and especially more recently in greening and sustainability design. Variable air volume systems (VAV), in particular, found a home during this era as they are more energy efficient than constant volume systems. An almost concomitant goal during this same period was the effort to lower HVAC equipment's first cost.

Since the late 1970s, there has been a growing interest in HVAC design to enhance indoor air quality. In particular, the focus has gradually shifted to increasing the volume of outside air introduced, the effective distribution of the air, and air cleaning procedures. ASHRAE has spearheaded action in this area with Standard 62, Ventilation for Acceptable Indoor Air Quality, serving as the construction and operating guidelines. Further, more advanced design concepts beyond the minimum code compliance of Standard 62 can be found in the IAQDG published by ASHRAE in 2009.

Design criteria for ventilation efficiency and effectiveness have been slow to emerge. Designers still tend to assume that air passing through diffusers will effectively reach occupants. Air flow patterns that "short circuit" the ventilation are still common place. Air cleaning technology for commercial and institutional buildings, for the most part, is still limited to low end furnace filters and pleated disposables.

Energy demands of the 1970s and early 1980s spawned financing schemes for capital retrofits, which forced engineers to become increasingly accountable for predicted energy savings. The whole performance contracting industry today rests on the engineer's ability to predict the amount of energy a given measure will save. Similarly, as IAQ becomes a high stakes game of productivity and lawsuits, engineers and architects of record will be held increasingly responsible for the indoor air quality the HVAC system delivers. This responsibility is apt to prompt more professionals to stay involved beyond the design stage (which could easily be worth an annual fee to the owners). Greater accountability and prolonged involvement will inevitably enhance HVAC design.

Integral to the design issue is the role of pressure relationships—between the outdoors and the indoors. This relationship is subtle—measured in low Pascals, meaning small increments on the water gage scale. For example, as little as 2.5 pa can drive infiltration into a conditioned building. This is uncontrolled and unconditioned air, which can negatively impact internal comfort conditions. In the case of external toxic pollutants, it is a source of exposure to occupants with potential fatalities. In the case of humidity control, this is a source of unconditioned air that can condense on vapor barriers, in building components, and on internal surfaces due to thermal bridges. This can bring about saturated substrates that become fungal growth sites that are, in many cases, concealed within the wall cavities or plenums. Improper pressure differentials can occur from a number of causes, including unbalanced supply and exhaust systems, stack effect within the building, external wind forces, and uncontrolled return air systems. However, most of the influential factors can be anticipated and accommodated at the design stage—or a building can be *designed to fail* if pressure relationships are ignored.

INSPECTING THE HVAC SYSTEM

It is amazing what can be found in the air-handling components of an HVAC system, if one only looks. Duct work is a disturbing example. If we could see the garbage heaps the air we breathe frequently pass through, we'd be appalled. The authors have frequently experienced ducts that have such heavy lint and dirt accumulation that airflow is impaired, turning vanes have been rendered inoperative, and control sensors are completely covered. It is normal to find the residue of construction—sawdust, wallboard dust, bits of wood and construction rubble—but we have also found the leftovers of lunches, including food and its wrappers, soda containers, and even lunch buckets. It is less normal but still much too frequent that we find the artifacts of animals—nests, feces, and even desiccated carcasses. These can run the gamut of birds, snakes and lizards, insects, bats and other rodents.

It is not surprising that buildings can make us sick! And that's only the duct work.

Many HVAC problems, such as obstructions in the outdoor air intake, can be caught by visual inspection. Just a little awareness and an educated eye can make a big difference. Other inspection procedures may require some simple measurements, such as the air flow at the diffusers, to see if the coil in a terminal box (terminal reheat) needs cleaning. HVAC inspections guidelines can be found in the "IAQ O&M Opportunities" listed under the specific components discussed later in the chapter. This has led the Standard 62 SSPC committee to incorporate a chapter in the current edition of the standard totally devoted to the subject of operations and maintenance even though a separate document, Standard 180, has been published.

HVAC OPERATIONS AND MAINTENANCE

It is virtually impossible to over-emphasize the importance of HVAC operations and maintenance (O&M). The inclination, is to first think of those components that come in direct contact with the air; e.g., filters and duct work. As important as that is, no part of the HVAC system can be ignored. The malfunction of any part of a HVAC system from worn bearings in a fan to dirty coil fins can affect the system's ability to provide quality indoor air.

There is much that HVAC technology can do to improve the quality of our air. Those opportunities seem more manageable if the component parts of a HVAC system are considered individually. Each of the section of the HVAC system provide rich opportunity for improvement of the indoor environment. Many of these opportunities will be revealed during the Phase II walk through discussed earlier in this book.

HVAC IAQ OPPORTUNITIES

The main purpose of mechanical ventilation is to provide a comfortable environment for the majority of its occupants. The IAQ focus makes sure it is a comfortable *and healthy* environment. To do this, the HVAC system provides many services including:

heating,	filtration,
cooling,	air distribution, and
outside air supply,	air diffusion.

A typical "HVAC" system schematic, shown in Figure 7-1, provides a way to identify the key points in a system before some variations are considered.

The main components of a mechanical ventilation system, as illustrated in Figure 7-1 are standard. These components, their graphic symbols and descriptions may assist non-technical personnel in interpreting the schematics shown in this chapter. These schematics and their presentation have been adapted from A Practical Maintenance Manual for Good Indoor Air Quality, which is an excellent manual for technician training. The symbols and descriptions follow Figure 7-1.

The "air side" of the HVAC system offers the most opportunities to improve the quality of indoor air and is the easiest to address. Unfortunately, air-side maintenance is also the most labor intensive.

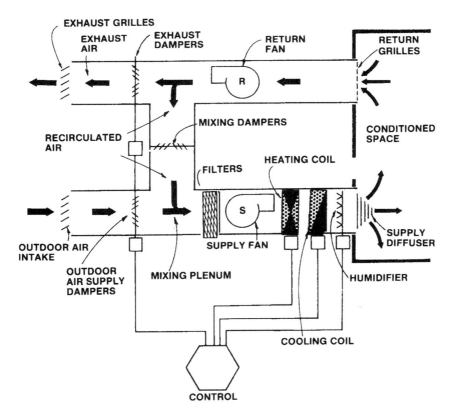

Figure 7-1. Mechanical Ventilation System

Controls The operation of the components are governed by controls to achieve the desired comfort end air quality in as energy efficient mode as possible.

Cooling Coils Only air-conditioning systems have cooling coils. They are also used to dehumidify the air during the summer season.

Dampers Exhaust, mixing, and outside dampers are designed to control the amount of air exhausted, admitted, recirculated and mixed. The dampers are linked and work in tandem.

Exhaust & Intake Grills/ Louvers Mounted on external walls, the grills and louvers allow for the discharge of return air and intake of outdoor air. Intakes and exhausts have "bird screens" to keep birds, rodents, snakes, etc. out of the system.

Filters Filters screen particles and bioaerosols from the air to protect the equipment downstream and clean the air we breathe.

Heating Coils One means of heating incoming air to adequate temperatures are heating coils.

Humidifier Water vapor or steam is injected into the air to increase relative humidity. They are usually needed only in winter.

Return Air Grill Air is removed from designated area through the return grill to ensure circulation.

Return Fan Return air is drawn from the rooms and pushed into the evacuation and mixing ducts by a return fan.

Supply   Conditioned air at the terminal box is delivered
Diffuser   to occupied areas by the supply diffuser to avoid
                                        drafts.

Supply Fan   The mixed air is blown through filters and across
                                        heating/cooling coils, into humidifier and finally
                                        to supply diffusers.

OUTDOOR AIR INTAKE

All air systems must have the capability to provide outdoor air to the building in order to replace air and dilute contaminants. The amount of outdoor air required varies with the type of facility and specific functions. For example, certain areas of a hospital, such as operating and delivery rooms, where contamination control is critical, have traditionally required continuous and total replacement of interior air with outdoor air. Research laboratories may also require total replacement.

The amount of outdoor air required to serve occupant needs is usually based on CO_2 content in the air. The traditional ASHRAE Ventilation guidance from Standard 62 is based on occupancy odor control that is a product of the bioeffluents of occupants. Since CO_2 is also a product of the occupants, it serves as an excellent surrogate for the other effluents. Earlier targets from prior versions of Standard 62, fixed the target for maximum CO_2 concentration at 1000 ppm. This represented a 700 ppm rise in concentration above normal outdoor air, which has a theoretical content of 300ppm of CO_2. A minimum ventilation rate of 15 cfm per occupant has proven to ventilate (or dilute) occupancy odors to an acceptable level while sustaining an internal concentration within the 700ppm band. More recent versions of the Standard have formalized the guidance to a 700 ppm rise above measured outdoor ambient (since polluted outdoor air can exceed 600 ppm). The 15 cfm per person minimum has persisted through several revisions of Standard 62, but changes in the 15 cfm determination although it prescribes higher levels dependent upon space usage and occupancy. However, the recent revision is more complex, as it recognizes the dual and additive contaminant contribution from both the occupants and the building. (See the discussion of Standard 62 later in the book chapter "What They Say.")

Outdoor air is drawn in, filtered, conditioned, passed through the space and exhausted from the building. The air intake location should allow good quality outdoor air to enter the facility and should be positioned so rain and snow do not enter the system. Just inside the air intake there should be a crude filter, a "bird screen," to prevent large objects from entering the system.

IAQ O&M OPPORTUNITIES:

— check intake location to avoid:

- reentry of building exhaust,

- odors and contamination coming in from garbage dumpsters/compactors,

- contaminants from bird droppings

- combustion contamination from indoor garages, loading docks, parking lots,

- exhaust from other buildings, and pollution from stagnant water;

— inspect bird screen to be sure it is free of obstruction at least twice a year (more frequently when intake is at ground level);
— be sure bird screen is intact;
— annually verify outdoor air locations and possible sources of contamination located near the intake;
— if there is any evidence of water penetration

- replace wet insulation with dry waterproof insulation, and

- install an indirect drain, change damper design if necessary;

— if occupants complain of stuffiness (or odors) check the quantity of outdoor air being admitted.

MIXING PLENUM

The outdoor air and the recirculated air are mixed in the mixing plenum. Any dirt in this area can be carried into the ventilation system.

IAQ O&M Opportunities
— check for any dirt, dust, critters or moisture; clean the mixing plenum as needed;
— seal against any leakage;
— verify that dampers are operative and correctly fixed; and
— verify that outdoor air is reaching plenum; clear air intake if warranted.

WATER AND AIR DISTRIBUTION SYSTEMS

Brief generic descriptions of water and air systems are offered to familiarize non-technical people with specific systems. These descriptions also provide a base of reference for the maintenance recommendations that follow. Those familiar with distribution systems may wish to turn directly to the discussion of O&M opportunities later in the chapter.

WATER SYSTEMS

Though many different air systems exist, there are basically only two types of water systems. While water system installation costs are usually higher, they are almost always more energy efficient and they avoid the contamination potential inherent in the long duct runs.

Unit Ventilator
In the unit ventilator, shown in Figure 7-2, the water moving through the coil is not controlled. However, the amount of air moving through the coil is controlled by the face and bypass damper. A thermostat and controller regulate the dampers. In this fashion, the air is cooled or treated as required.

In addition to the face and bypass damper, the unit ventilator generally has two other dampers: the fresh air damper and the return air damper. These dampers may be set in a fixed position to allow a constant percentage of outdoor air to enter the room. They may also be controlled by a thermostat to allow for variable outdoor air percentages, depending upon the control scheme selected. Some unit ventilators have controls that allow for 100 percent outdoor air.

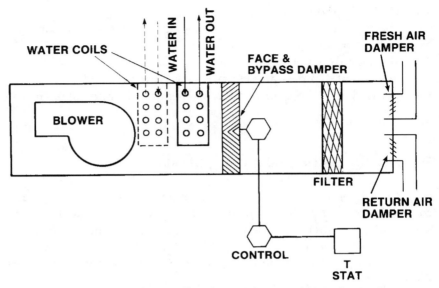

Figure 7-2. Unit Ventilator; No Water Volume Control

Fan Coil Unit

The second basic type of air moving unit is the fan coil unit as shown in Figure 7-3. The fan coil unit is the simplest of the various systems available for air movement and control. There are no dampers or controls inside the unit.

Control of the fan coil unit is by means of a water valve which reacts directly to the thermostat setting. The control of water through the coil is the only built-in control for the temperature of the leaving air. Blower motor speed control is not commonly used. Fresh air is usually controlled by a damper in a fresh air duct.

The thermostat, *not* the air moving unit, should control the temperature of the air being delivered to the room through the unit ventilator or the fan coil. Adjustments to a thermostat should be made in very small increments with enough time allowed after each adjustment for the change to have an effect on the space temperature. Too great or too rapid a change in thermostat settings will frequently lead to over-correction of the space temperature.

Water Source Heat Pump

Heat pumps can pump heat out of a space or into a space (to or

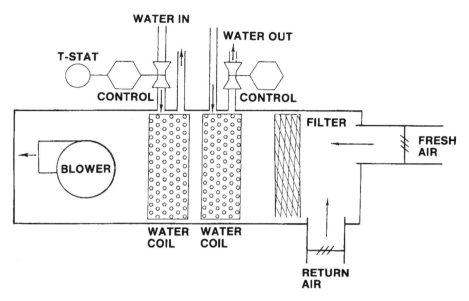

Figure 7-3. Fan Coil Unit

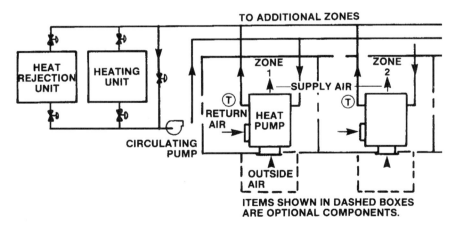

Figure 7-4. Unitary Hydronic Heat Pump System

from the outdoors). Water source heat pumps have a water loop inter-jected between the pumps and the outdoors. Heat is pumped to or from this circulating water loop.

AIR SYSTEMS

While air systems generally use chilled water from a central plant, they are sometimes outfitted with D-X or heat pump coils. There are many types of air systems. They fall in one of two categories: constant volume and variable air volume. The "constant volume" systems require a fixed amount of air to be distributed at all times. In the most common of these units, single and multizone, the duct work can be sized large enough to keep the air velocity (speed) low, resulting in a relatively low fan operating cost. Constant volume systems vary temperature to satisfy different load conditions.

Single Zone

In constant volume systems, air is moved to one or more locations via ducts. A single thermostat usually controls the amount of chilled or hot water going to the coil, or turns off/on electric heaters and turns off/on a refrigerant valve to the coil if D-X cooling is used. Normally, the fan always operates at the same flow rate. Auditoriums, dining halls, gymnasium and other large open spaces are usually equipped with ducted single zone systems.

Single Zone with Terminal Reheat

An adaptation of the single zone system can be made with the addition of zone reheat coils and can offer some of the same advantages as a multizone system.

Multizone

The multizone air handler has one constant volume fan which supplies heated and cooled air to several individually controlled (thermostats) zones. The multizone system includes both the heating and cooling coils within the air handling unit (Figure 7-6). The multizone system controls different zone requirements by varying the amount of hot air and cold air to the zone. Cooled air is produced at constant temperature and then mixed with hot air to meet the room temperature required by the individual zone thermostat. Although this type of system can provide the consistent temperature desirable for indoor air quality, it is a costly process.

Dual Duct

The dual duct system is similar to the multizone system in that it has constant-volume air supply and a hot and cold deck. In this system,

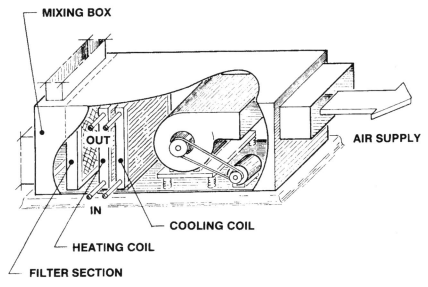

Figure 7-5. Constant Volume, Single Zone

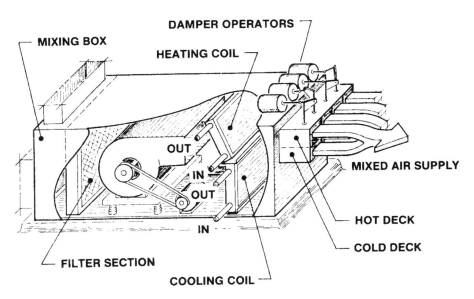

Figure 7-6. Multizone System

however, air from the hot deck is ducted to each zone and air from the cold deck is ducted to the same zone. At each zone, in the plenum space, is a mixing box where the two air streams merge to provide tempered air according to demand from the thermostat.

Although this procedure provides excellent temperature control and a comfortable internal environment, it is hard on the utility bill. The duplication of duct work also presents some IAQ maintenance headaches, as a separate set of two duct runs is needed for each space.

Cooled air is mixed with heated air after both streams are ducted to each zone. Medium to high pressure ducts are used because the flow of air in each duct varies from 1 to 100 percent of the required air flow for each space.

All constant volume systems share the same generic drawback—they move a fixed quantity of air at all times regardless of the actual building load requirements. In so doing, each of these systems (except the single zone unit) must operate on a "reheat" basis in which adding heat to the airstream is the only way to produce comfortable room

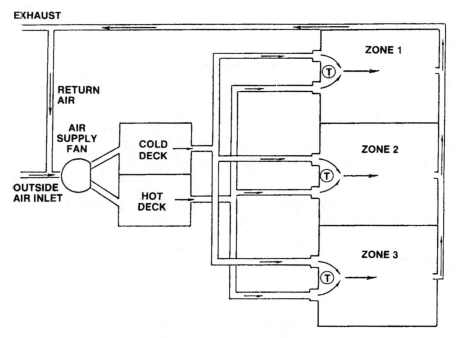

Figure 7-7. Dual Duct System

conditions. As a result, constant volume systems must be designed large enough to handle the greatest load which may be required by all building zones simultaneously. The system's first cost is high since the equipment must be larger than is normally needed, and operating costs are likewise excessive.

The development of the "variable volume" air distribution system has provided designers a more efficient means of distributing comfort heating and cooling to building zones. This system is designed to meet all building load requirements while, at the same time, allowing energy savings at the fan and the heating/cooling plant.

Variable Air Volume

The variable air volume (VAV) system maintains temperature in a space by controlling the amount of heated or cooled air flowing to it. Conditioned air is delivered at medium to high pressure to the room through "VAV boxes" containing dampers which restrict the flow as the room thermostat dictates.

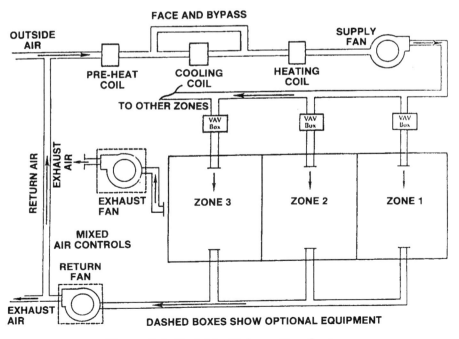

Figure 7-8. Variable Volume Fan System

As a VAV system modulates from its cooling design level (maximum) to its heating design (minimum), total air circulation rate reduces. Since outdoor air is admitted as a constant percentage of the total air circulated, the outdoor air intake is reduced proportionately. Many VAV systems, however, have economizer cycle controls.

In a VAV system, the amount of outdoor air admitted at any one time depends upon:

(a)   the total amount of air circulated (to which the percentage is applied); and

(b)   the percentage of outdoor air brought in by the economizer cycle control.

If room temperature conditions are satisfied, the VAV could theoretically shut off the flow of air. However, health and safety reasons dictated, even during the energy crunch, that a minimum flow of air be maintained. VAV systems generally employ VAV boxes and medium pressure ducts. Since the late 1970s constant volume/low pressure duct systems have been retrofitted to have VAV characteristics.

•   *Ceiling Bypass VAV*—Supply ducts to a zone are outfitted with damper boxes that "dump" supply air to the return air plenum as cooling becomes unnecessary. Thus, the air flow through the system is constant. The damper gradually varies the flow to the room or to the plenum as needed.

•   *Variable Volume Single Zone*—instead of reheating cooled air in a single zone system, the air supply is varied by a variable speed drive or variable inlet valves in response to a pressure sensor located in the duct work.

•   *Side-pocket Fan VAV System*—This augments the ceiling bypass or VAV single zone by providing a fan at the duct terminal. This fan draws air from the plenum and forces it through the register, along with the cooled air, into the zone. This maintains air circulation and air distribution patterns in the space.

All VAV systems as well as these hybrids govern air changes in response to temperature needs. Only indirectly do they respond to the number of occupants in a room; therefore, VAV systems are not as responsive to CO_2 levels or other contaminants.

In a facility without operable windows, a VAV system's effective ventilation rate to a number of spaces can vary radically. This is particularly true when indoor and outdoor ambient temperature are very close.

Since the VAV characteristic of throttling back outdoor air as a temperature response may result in unacceptable lower limits, some method to set a floor in outdoor air is needed. This requirement can be met by supplementing the VAV system with a constant-volume air supply in combination with the VAV system or independently. What would seem to be an obvious alternative, presetting for a minimum total air quantity could result in overcooling. Putting the supply air temperature at a higher value would increase total air flow, but at a cost of more energy for fan power.

Operations and maintenance opportunities related to key components, such as dampers, coils, fans, filters, are common to all the distribution systems and will be treated by component parts.

Fans

 Indoor air quality is dependent on a fan to move air through the system. If fan blades are dirty or the belts are worn or stretched, fans will not work as effectively as they should and or may not run at their normal speed. When the static pressure and air flow are reduced, the air may not reach the last terminal box and areas at the "end of the line" do not receive sufficient air exchange.

IAQ O&M Opportunities:
— check motor amps and volts with the fan running;
— listen for motor coast down noise;
— measure static pressure across the fan and check it for excessive vibration;
— determine if the last terminal box in the distribution system is receiving sufficient air flow;
— change fan rotation speed if warranted;
— clean blades with steam or a solvent annually;
— check the belts and pulley grooves for signs of wear or damage; and
— verify that belt tension is adequate; tighten or replace belts as needed.

Bearings
— inspect and lubricate fan and motor bearings;
— make sure grease line connections and motor mounts are tight; and
— with fan in operation, make sure the shaft revolutions do not exceed the maximum.

Dampers

Dampers affect the amount of air intake, the amount recirculated and the amount exhausted. When dampers are not operating properly or are loose, air flow is not controlled and there may be unacceptable swings in temperature. In addition, the amount of "fresh" air may not be sufficient to assure indoor air quality. Conversely, improperly operating dampers may bring in too much air, which will not help IAQ and could require costly conditioning of unnecessary air.

IAQ O&M Opportunities:
— check linkages, tightness and operation [Remember dampers are located throughout the system -outdoor and return air systems, bypass ducts];
— make sure damper actuators are working properly;
— repair, adjust and seal dampers as needed.

Heating and Cooling Coils

Heating coils warm the air to assure comfortable conditions for the occupants. If dirt obstructs the coil or the controls are not operating correctly, temperatures may be unacceptable. Control problems may stem from the design of the coil, a heating valve that is not functioning correctly, or a faulty or poorly adjusted control system.

Cooling coils cool and dehumidify the air supply. Since the coil dehumidifies the air, water gathers. In addition to heating coil problems, the cooling coil must be drained properly to avoid biological growth.

IAQ O&M Opportunities:
— measure the heating coil discharge temperature regularly; if too high or too low, the control system needs to be checked;

— check if the water or steam is being closed off but air is still being heated. If this is occurring, the valve is not closing tightly;
— inspect and clean entire coils periodically; be sure coil fins are clean;
— make sure the off season cooling coils are "winterized;" and
— be sure condensate drains and drain pans are clean and operating properly, without pooling or retained water.

Humidifiers

The humidifier supplies humidity to the air. The amount of humidity required is directly related to the quantity of outdoor air being brought into the facility, the degree to which the air must be warmed and the relative humidity (Rh) of the outdoor air. Humidifier IAQ problems include stagnant water and water carryover into duct insulation. The prevalence of humidifier fever attests to the importance of cleaning and maintaining humidifier reservoirs.

IAQ O&M Opportunities:

— check humidifier sprays, grids and pans for scale buildup and plugging;
— adjust the steam float valve;
— calibrate and verify the humidity sensor frequently;
— empty and clean pans when humidifier is not operating;
— make sure downstream duct insulation is not wet; if wet, replace; and
— check for proper operation of humidifier controls, including high limit humidistat on controller.
— assure that overflow drain lines are free-flowing, and clear of debris or algae buildup.

Filters

In addition to the primary role filters serve in removing particles and fibers from the air, they reduce the microorganisms that attach themselves to these particles. Filters also prevent dust and dirt from accumulating on work surfaces, walls and equipment. This reduces custodial work and helps maintain equipment. It also relieves occupant concerns since cleanliness is usually associated with air quality. Filters also preserve the "health" of the entire ventilation system by preventing the accumulation of material downstream in ducts, fans coils and other ventilation components. It is far more difficult and costly to fix downstream problems and to remove contaminates than to maintain an effective filtration system.

Air cleaner locations may vary with the purpose of the filter. If the contaminants are primarily from the occupied space, then it is logical to place the filter in the recirculated airstream. If filterable contaminants from outdoors are a concern, then filters can be placed in the make-up or mixed air. Since outdoor air has been described as a natural reservoir for fungi and many occupants have reactions to pollen, the growing consensus is that outdoor air as well as recirculated air should be filtered.

Media filters and electrostatic air cleaners are typically found in building HVAC systems. Neither will function well without regular maintenance. As particles build up on the media filter, it will increase collection efficiency and screen out smaller particles. This efficiency, however, does not justify leaving media filters unattended. As particles build up, filters become clogged and the critical air flow rate decreases. Clogged filters increase resistance and reduce system efficiency, while increasing energy usage.

Electrostatic air cleaners (EAC) will not operate properly unless kept clean. Build up on the charged plates interferes with the unit's effectiveness and, in fact, dirt on plates acts as pollutant sources, especially odors.

Design considerations that effect filter maintenance include:
- Filter efficiency is not correctly selected for the contamination load. Or for reasons of cost, filters may have been downgraded.
- Filter accessibility. Location of the unit, limited space, doors that are obstructed or held in place with screws that discourage filter changes.
- Filters that don't fit the opening. Filters that don't cover the opening cause a bypass into the open area where there is less resistance. Or, the gaps may be covered with other material, such as plywood. This can increase air velocity across filters, which, in turn, can increase static pressures.
- Filter seal is a further place that air can by-pass. Gaskets fail, erode, or fall out leaving gaps around filter cartridges. Filter frames that are made of beverage board or cardboard can dis-

tort and bow allowing by-pass in side-load slide-in retainers. Filter frames may not be properly gasketed or caulked allowing by-pass between or around the retainer.
- Filter gages are uninstalled or malfunctioning, which does not enable the filter bank to be serviced on the most cost effective change cycle.

Unless there is special filter expertise on staff, filters should be selected by specialists, who understand the problems and the options available. General working knowledge by practitioners who are not marketers of a particular brand of filter is sparse. Thus, as an aid to building owners and managers, this book contains a chapter dedicated to air filtration with more technical detail than normally found in this type of text.

IAQ O&M Opportunities:
— inspect and change filters regularly. Filter changes may be periodically scheduled or done best in response to static pressure measurements;

— check and record static pressure loss at the filters regularly; when pressure reaches manufacturer/design level, or a preset point, the filters must be changed. (Be sure static pressure indicators are working properly.);

— stop ventilation system to change filters wherever feasible to avoid re-entrainment of particulates; and

— remove expended filters from the conditioned space or mechanical room to avoid re-entrainment of odor.

Ducts

Air can travel literally miles and miles through duct work. Poorly maintained duct work can add pollution. (Dust splays or accumulation near the diffusers may not be from upstream duct work but charged particles from the room.) Moisture in the duct work encourages microbiological growth, regardless of the duct construction material.

The North American Insulation Manufacturers' Association (NAIMA) advises that fibrous glass lined duct work can be cleaned if the problem is an accumulation of dust and spores. It cannot be

effectively cleaned if mold growth within the fibrous itself has occurred. Fibrous lining should not be used in areas of high humidity
or near moisture sources. Work by Tulis and Thumann of Duke
University indicates the risk areas for moisture accumulation and
subsequent fungal growth occurs primarily where hot and cold air
co-mingle in the distribution system. Their survey of buildings on
Duke's campus revealed that mold seldom was found even in high
humidity areas like outdoor air plenums, unless thermal shock had
occurred.

IAQ O&M Opportunities:

To reduce the probability of mold growth in duct systems,
SMACNA (Sheet Metal and Air-Conditioning Contractors National Association, Inc.) in its publication, *Indoor Air Quality*
makes the following system design and maintenance recommendations:

— promptly detect and permanently repair all areas where water
   collection or leakage has occurred;

— maintain relative humidity at less than 60 percent average in
   all occupied spaces and low velocity air plenums. During the
   summer, cooling coils should be run at a low enough temperature to properly dehumidify conditioned air;

— check for, correct, and prevent accumulation of stagnant water
   under cooling deck coils of air-handling units, through proper
   inclination and continuous drainage of drain pans;

— use only direct generated steam as the moisture source for
   humidifiers in the ventilation systems. Boiler steam should not
   be used because it can be contaminated with volatile amines
   (used as rust inhibitors);

— once contamination has occurred (through dust or dirt accumulation or moisture-related problems) downstream of heat
   exchange components (as in duct work or plenum), additional
   filtration downstream may be necessary before air is introduced into occupied areas;

— air handling units should be constructed so that equipment
   maintenance personnel have easy and direct access to both
   heat exchange components and drain pans for checking drainage and cleaning. Access panels or doors should be installed
   where needed; and

— non-porous surfaces where moisture collection has promoted microbial growth (e.g., drain pans, cooling coils) should be cleaned and disinfected with detergents, chlorine-generating slimicides (bleach), and/or proprietary biocides. Care should be taken to insure that these cleaners are removed before air handling units are reactivated.

Fibrous glass insulation that has become wet in service should be removed and replaced to reduce the risk of mold growth, and to restore thermal and acoustical performance levels. Otherwise, this insulation may be very difficult to dry *in situ* under normal operating conditions.

The cleaning of duct work, whether lined or not, is a difficult task, and may have dubious value or in some cases negative impact. Research performed by the Florida International University on residential properties in southern Florida resulted in elevated contaminant levels, both particulate and fungal, in homes subsequent to having duct cleaning performed. Flowing large quantities of air through the system is not generally effective for anything but large pieces of extraneous material in the system. This is due to the fact that air boundary layers which form at the duct surface are at very low velocity. This makes the entrainment of dust particles in the air stream difficult, if not impossible. Systems should be carefully evaluated before assuming that duct cleaning will automatically increase air quality. Further, dirty air conveyance systems are usually a symptom of dirty supply air. Thus, cleaning should not be undertaken without an understanding and remedial action regarding the root cause of the condition. NAIMA has published a guidance document on "Cleaning Fibrous Glass insulated Air Duct Systems" and EPA has issued a bulletin resulting from their research on cleaning residential ducts. Both organizations urge caution when determining the need and selecting contractors to perform the duct cleaning.

Terminal Boxes/Diffusers

The terminal box is the last step in air flow to the room. It reduces the pressures in the system to the room's atmospheric pressure. If the mechanism is jammed, air flow is restricted. Whether the terminal box uses a metal register, perforated grillwork or a valve, regular maintenance is required.

IAQ O&M Opportunities:
- verify operation periodically;
- clean terminal boxes; inspect and on a regular PM program;
- when heating is at the terminal box, check air flow at diffusers; and
- if room is perennially not warm enough on a dual duct system, leakage may be occurring. If leak exceeds 10 percent, clean and replace joint gaskets. If off by much more than 10 percent, clean-finned coils and make sure any heating coil valve is not leaking.

Heating/Cooling Plants

The heating and cooling plants themselves present some concerns not addressed in the discussion of components.

Heating Plants

Heating plants always have a level of incomplete combustion, which produces combustion contaminants. The level of contaminants generated depend on the combustion characteristics of the burners and furnace, the type of fuel used, and the burner operating mode. Leaks allow these contaminants to escape and ultimately reach the occupants. Burner/fuel ratio is a key operating parameter that affects combustion products generated, the build up of deposits internally, and the quantity of smoke emitted from the stack. Boiler stack emissions may also enter the building through air intake cross contamination.

IAQ O&M Opportunities:

- check heating plants in the fall and during the heating season;

- seal cracks; and

- check burner settings and make sure burner parts have not deteriorated to avoid high CO emissions.

Cooling Plants

Cooling plants that have elevated chilled water temperatures can contribute to indoor air problems. Adjusting chilled water temperatures upward as the load decreases can push relative humidity up to 65-80 or higher percent in occupied spaces. This not only makes the occupants uncomfortable, but increases the likelihood of microbial growth. It has a particularly disastrous effect because humidity control is lost while start-up and recovery times are delayed.

Special care should be taken with cooling towers that have been dormant or deactivated as these are near perfect sources of Legionella exposure to workers that performing maintenance or replacement.

PAY NOW OR PAY LATER

All too often the "INDOOR AIR PROBLEM!" hysteria has a typical pattern. Early complaints are ignored. Management only recognizes it has a problem after the staff is upset, absenteeism grows, productivity drops and employee emotions run high. Owners only recognize they have a problem after tenants move out and the space remains empty. Many times neglect, deferred maintenance, and deteriorating performance of the HVAC system is a contributor to these outcomes.

Faced with economic losses and damaged relationships plus the pall of threatened lawsuits, the pressure is on to find the culprit. In a quandary, owner/managers rush out to hire a consultant and the big hunt for the contaminating villain and its source begins. Suddenly money becomes available when the 11 o'clock news anchor is at the front door. Unfortunately, the few thousand dollars of cash savings from deferred maintenance will became millions in losses of revenue, productivity, litigation, and damaged public image. The lesson to be learned is the classic "pennies now—dollars later" outcome.

Chapter 8

FILTRATION:
THE UNDERUTILIZED IAQ ASSET

INTRODUCTION

Air filtration is emerging as an important element in indoor air quality (IAQ). It is often involved with other factors which create or allow inadequate IAQ. Further, it is a routine recommendation for completing the mitigation process when space must be cleaned up after IAQ-related problems have occurred. Because of this essential aspect of this often misapplied and under-used technology, this chapter will discuss air filtration and air-cleaning from the perspective of indoor environment in commercial buildings. The discussion will focus on how both particulate and gas-cleaning equipment work, their background, where they are to be applied, their use to comply with ASHRAE Standard 62, and their contribution to energy management, sustainability, efficient building operations, and building security. First, let us discuss the history of filters.

HISTORY OF FILTERS

Air filtration is perhaps misunderstood by many designers, ignored by some, and forgotten or dismissed by many others. As a result, filtration and its benefit (clean air) is too often overlooked in planning for adequate IAQ. Poor filtration in both residential and commercial buildings over the years has created an entirely new industry called "duct cleaning." Unwanted contaminant buildup (whether in space, ductwork or air-handling unit (AHU) can be a contributor to sick-building-related problems. It can promote microbial growth; expose occupants to respirable particles; and impair the performance of heating, ventilating and air-conditioning (HVAC) system components.

Although the commercial building construction has largely overlooked filtration, American industry has relied heavily on clean air to enable the robust growth of technology since the World War II. Ultra high-efficiency filtration is a relatively old and well-developed technology. It has been over a half century since the development of high efficiency particulate arrestor (HEPA), as it was originally designated. More recently, the term "HEPA" is defined as a High Efficiency Particulate Air filter. The importance of this development was that it established 0.3 micron as the most penetrating particle for arrestance-type filters. This has remained as a standard for determining filter performance and has laid the groundwork for 99.97% or virtually "absolute" filtration efficiency. It also sets the fabrication pattern of deep pleated filter media as the manufacturing style of high-efficiency filters.

HEPA filters were developed during the World War II as part of the Manhattan project to control a particle determined to be radioactive iodine. Iodine has the unique ability of sublimating or jumping from gaseous state to solid state without going through the liquid state. Thus, radioactive iodine appeared around nuclear reactors as a dangerous toxic particle in the form of an airborne condensation nuclei of about 0.3 micron in size. The HEPA filter was developed to protect the environment and the workers from this potential radioactive exposure. The product was later commercialized under patent and marketed under the registered brand name of "Absolute" filter by the Cambridge Filter Company who acquired the patents and initially marketed the HEPA filter. The wide acceptance and usage of this brand name by the filtration industry risked it going generic as had "nylon." This forced the creation of a new acronym to serve as the generic noun for this class of filter, thus, the birth of the High Efficiency Particulate Arrestor (HEPA). In the 1960s, the technology patents were legally challenged and become public domain to be pursued competitively by other American and international firms.

By the 1960s, hospitals were protecting critical spaces such as operating suites and OB/GYN rooms with extended media bag filters and/or electrostatic precipitators now called electronic air cleaners [EACs]. These filters were used to create relatively pathogen-free supply air in hospitals and their usage in healthcare facilities was promoted and enabled by Federal legislation that underwrote the financing of hospital expansion in the 1970s under the Hill-Burton Act.

The trip to the moon in 1969 was made possible by huge clean rooms where giant computers were built to manage the elaborate flight planning. Later in the space program, special gaseous and particulate filtration units aboard the Lunar Exploratory Module (LEM) allowed the manned exploration of the moon's surface.

Special gaseous filtration media and systems were developed to protect computers and process control apparatus from the gases and contaminants present in hostile industrial settings. Although early systems relied on an activated carbon, this application area was soon dominated by potassium permanganate impregnated alumina. The use of these high-efficiency filtration systems integrated throughout the electronics industry and became the standard for fabrication areas for delicate electronic chips and computer components. These technologies were also used in the pharmaceutical and medical products industries to control corrosive or reactive gases.

There is a long history of highly effective filtration that has been predominantly focused on contaminant control in specialized or industrial spaces. Until recently, designers and suppliers of typical HVAC systems in commercial buildings were using disposable panel (throwaway) filters that were developed in the 1930s as furnace filters. Furnace filters were developed and used in central hot-air furnaces for the sole purpose of keeping the fire box in the unit from catching fire from flammable accumulations of household lint, animal hair and other fibers. This type of heating system gained popularity over hydronic or radiant-type heating system.

In the 1950s and 1960s, hot-air furnaces and their blowers were essential components of the air conditioners. Furnace filters were cheap, thin (usually no more than one inch in media thickness), had low airflow pressure-drop, and were disposable. Early versions were made of fabric, metal wire screening, or animal hair mats. In later 1930s, they were made of the newly developed man-made spun glass fibers called fiberglass, and the filters were called "duststops." In commercial applications, they were increased to two inches of fibrous glass matt that provided higher dirt holding capacity but did not increase efficiency significantly.

Later developments of low-efficiency (refer to Figure 8-1) panel filters have simply built on the original fibrous matt technology. Fibers were varying to a wide range of lofted natural and synthetic materials. Permanent metal frames and replaceable media pads attained popularity. The media were then mounted on rolls in elaborate machines which automatically advanced as the media loaded. Synthetic polymer media was applied in blankets, or fabricated into wire-reinforced panels using heat sealing techniques. The denier or density of the media was varied and/or combined and various thicknesses were made available. These developments altered and enhanced their dirt-holding, efficiency and airflow characteristics up to a certain degree. This media selection usually was highly influenced by available by-products or wastes from the fabric and batting manufacturers. Recent modifications have employed the inherent or an imposed electronic charge on the polymeric-type fiber mattes to enhance the particle collection and retention properties of filters using such synthetic filter media.

Meanwhile, extended media filters were being developed to serve the needs of specialized spaces. These would include specialized commercial spaces which experience high lint loads such as textile manufacturing plants, retail department stores, marketing fabric products, medical and health-care facilities, and large public assembly buildings. The most compelling marketplace that emerged during this period was the medical and health-care market. In the 1960s and early 1970s, as part of government grant legislation, high-efficiency air cleaning attained wide use in critical care areas such as operating rooms and OB/GYN floors. The electrostatic precipitator air cleaner attained early acceptance for this area of application.

"Extended media filter," as shown in Figure 8-2, refers to enhanced surface area of the filtering medium. This was done by pleating or corrugating the media in shallow, and subsequently, in deeper pleated pockets or extensions. Other tactics included shaping the media in conical, cylindrical, cubical, wedge or rhomboid forms that increased the surface area of the filter within the given frame size of the filter. The advantage of this extended surface was that the velocity of the air could be reduced, therefore reducing the airflow resistance. This enabled the use of denser filter mattes that provided higher efficiency and the retention of smaller size particles. The extended filter surface also yielded higher dirt holding capacity and resulted in longer useful filter life.

Earlier versions employed elaborate metal wire or mesh frames to support the filter medium that was either porous cellulose or natural fiber mattes. As fabrication technologies improved and new filter media were developed, non-supported versions were developed (bag or pocket filters) (refer to Figure 8-3). Highly specialized fabrication techniques resulted in pockets that were sewn with progressive sized stitches, face plates with pockets glued to the metal surface, etc. In a shallow pleated filter, more recent high speed corrugating and fabrication technologies

Figure 8-1. Typical Low Efficiency Filters (*Courtesy of Filtration Group*).

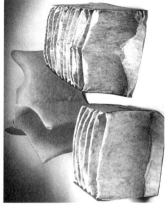

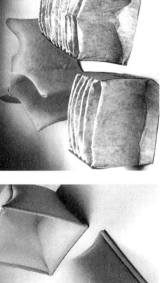

Figure 8-2. Typical Extended Media Filter Types (*Courtesy of Filtration Group*).

allow the mass production of one-, two- and four-inch pleated filters for medium-efficiency filtration applications. The latter pleated filter has attained wide spread acceptance and application in commercial building HVAC and is making inroads into the consumer market though DYI outlets.

An early version of the extended media filter developed in the late 1930s employed a porous layered cellulose crepe paper type media. This was pleated into a metal mesh filter retainer system using a manual corrugation machine located on site (PL-24). This led to the prefabricated cartridge which mounted on an elaborate wire support cage that held the filter material rigidly in the airstream (HP). With the development of a superfine fibrous glass filter matt, manufacturers were able to develop filters having much higher efficiency than previously possible.

A more recent development of extended media filters is the combining of pleating, surface extension and HEPA fabrication. This is best represented by a high-efficiency extended media filter which uses corrugation, or "mini-pleating" of HEPA type filter media (refer to Figure 8-3). This is then shaped into a wedge-shaped cartridge that provides even more ultimate surface area. The resulting module provides high efficiency combined with high air flow and long service life.

APPLYING FILTRATION

Air-cleaning and filtration technologies can be widely used in modern commercial buildings. The following lists the areas of interest to the designer and facility manager:

- Protect mechanical equipment from buildup of energy-consuming films and layers;

- Protect occupied space from unsightly dirt splays and dust accumulation;

- Protect occupants from internally generated gaseous chemicals that cause irritation or more severe toxicity;

- Protect occupants from irritating or harmful respirable particles;

- Protect processes from contamination-caused rejects;

- Provide clean make-up air free of external pollutants;

Figure 8-3. Typical High Efficiency Filter Types (*Courtesy of Filtration Group*).

Figure 8-4. Typical HEPA Type Filters (*Courtesy of Filtration Group*).

- Protect environment from contaminated building exhaust;

- Provide source control to spot treat filter problem space or sources;

- Augment ventilation by treating return air for use as equivalent fresh air; and

- Enhance the protection of occupants against airborne contaminants of mass destruction (WMD) and toxic industrial chemicals (TIC).

The following list outlines some of the potential cost and performance benefits of the appropriately used filtration:

- Increase system efficiency with maintained high levels of heat exchange efficacy;

- Increase system life with reduced wear and contamination;

- Lower maintenance costs by avoiding premature cleaning and equipment failures;

- Lower housekeeping costs with reduced contaminant loads in a conditioned space;

- Avoid product failure due to contamination during the manufacturing process;

- Increased productivity of personnel through reduced absenteeism;

- Reduced energy consumption and enhanced sustainability because of system cleanliness and heightened efficiency; and

- Reduced health risk of occupants due to reduction of irritating, pathogenic, viable, or toxic chemicals.

BASIC PRINCIPLES OF AIR CLEANING

Airborne contaminants are both gaseous and particulate. They vary drastically in size as they are carried in an airstream. It is important to understand the size of particulate matter as this is critical to the behavior of filters in controlling that specific size fraction. Figure 8-5 shows the size of common airborne components. Note that pollen is relatively large with a mean particle diameter of approximately 40 microns to 50 microns. A micron or micrometer is 1/25,400 of an inch. This can be compared to the even larger human hair that averages around 80-100 microns in size. Conversely, the size of airborne fungal spores range from 2 microns to 12 microns with a mean particle diameter of approximately 3 to 5 microns. Tobacco smoke is even smaller and is in the submicron range as it is a by-product of combustion and starts out as a condensation nuclei. Gaseous matter is even smaller since these contaminants are carried in solution at the molecular level in an airstream. They are measured in Angstroms equal to one ten-billionth of a meter.

In designing filtration systems, designers should be able to differentiate between efficiency and efficacy. Efficiency is the comparison of incoming quantity to outgoing quantity. For example, a brick wall is a highly efficient filter. Yet, because it will not allow air flow easily, it is not an effective filtering device. Efficacy, on the other hand, includes consideration of the desired result and is a better way of assessing the total performance of an air-cleaning system. To evaluate and to properly specify and apply filtration equipment, a designer must consider other factors besides efficiency. These include factors that influence the overall efficacy, such as:

- Life cycle;
- Capacity;
- Air flow characteristics;
- Efficiency against the contaminant of concern;
- Initial cost;
- Service life cost/unit time;
- Labor of installation and replacement; and
- Disposal/recycling cost.

Life Cycle Cost Assessment (LCCA), widely used in the Green Building movement to express the entire ecological cost impact of a product, is also applicable here. The analysis of the real cost of a filtration system must, therefore, include in addition to its initial cost, its efficacy, labor, life cycle, energy demand, as well as the ecological effect from cradle to grave. The latter may take the form of raw material depletion, environmental effect from manufacturing, cost of solid waste disposal and the inherent energy burden. In certain locations where solid waste disposal is a major issue, filters must be completely disassembled into their components to facilitate recycling.

The bottom line is that life cycle cost of filtration and air cleaning is dominated by the energy cost and the installed life cycle. Thus, the best value, regardless of initial filter cartridge price is the filter having the required efficiency that requires the least pressure drop and lasts the longest. This combination produces the lowest energy usage and the

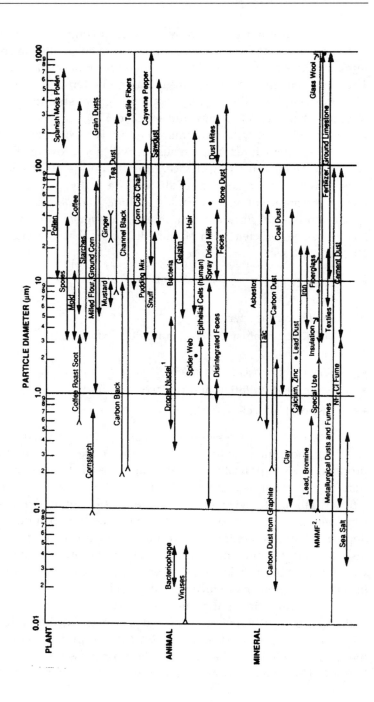

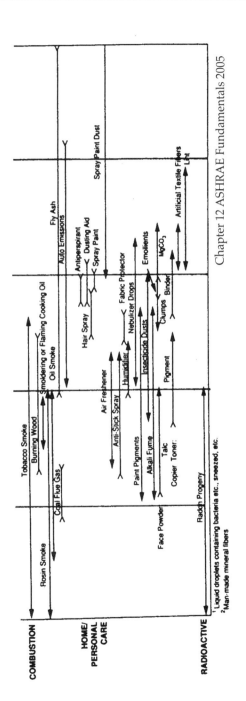

Chapter 12 ASHRAE Fundamentals 2005

Figure 8-5. Size Distribution of Common Airborne Contaminants.

least handling and disposal cost and will yield the lowest cost per unit time of installation. Thus, life cycle analysis is required to determine the best overall performing filter system selection from a value standpoint.

PARTICULATE FILTRATION

When detailed performance characteristics and application technology are discussed, the specific type of filtration, either gaseous or particulate must be defined. This is because the filtration capture techniques and physical forces vary drastically. The control technology of particulate matter are:

• Impingement
• Straining
• Electromagnetic attraction
• Diffusion

Impingement implies a "collision." For example, a particle will impinge (or collide) with a filter fiber and, once intercepted, it will tend to cling to that site (refer to Figure 8-6). This is the predominant mechanism for large particle filtration such as paint booth filtration, as well as the whole range of low-efficiency prefilters. The performance charts will also demonstrate this phenomenon with very small particles less that 0.3 micron in size.

Straining results when the filtration media forms small holes or paths for air to flow through (refer to Figure 8-6). Particles larger than

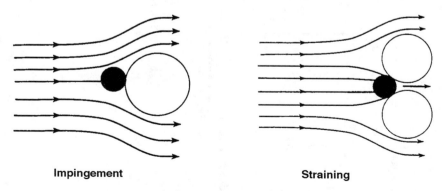

Impingement   Straining

Figure 8-6. Typical Methods of Particle Control.

that path will be strained out and then retained within the media. Denser media with smaller air passages will be more effective against smaller particles. Particulate filters relying on either impingement or straining are generically called mechanical filters.

Electromagnetic fields are used to actively alter and manipulate the electrical charge of the airborne particle enabling it to groundout onto a collecting plate or media. The particles are then held into place with electromagnetic forces. Filtering devices relying on an electrical power source and this capture technique are called EAC (electronic air cleaners).

Diffusion reflects the property of small particles to remain in motion independent of the air stream as they seek equilibrium and respond to minute electron charges and activity with other molecule and superfine particles in solution in air. This activity enhances the probability of impingement with fibers of the filter media and increases likelihood of retention of the smaller airborne particles.

Various media that are used in mechanical filters include: (a) paper matte, usually in the form of a felt-like matte; (b) natural fibers, such as cotton and animal hair; (c) man-made fibers such as superfine fibrous glass; and (d) a variety of synthetics in the form of a lofted bat.

Certain of the synthetic fibers have an inherent surface charge or can be altered to enhance their natural charge, referred to as "electret" media. In addition to simple straining, this allows the media to employ an electrostatic charge to capture particles. Designers need to be cautioned that electrostatically charged media can lose their charge during their life cycle. This causes them to revert into strainer filters which can deteriorate particulate efficiency and retention. This can also be detrimental to their performance and harmful to the airstream.

Efficiency is not always predictable. The factors affecting the particle behavior include its size, mass and charge. In fact, the efficiency of a filter highly depends on particle size. This is why a rating system that does not specify the precise size of the challenge particle may be misleading. Furthermore, the type and loading characteristics of the filter media or device will also affect efficiency and ultimate performance. Strainer media such as that used in bag filters will increase in efficiency as it loads. Thus, a rating system that uses averaging over the life cycle of the filter may provide misleading information on the efficiency of the filter particularly during the early portion of its life. Conversely, EAC filters lose efficiency over time as the electromagnetic plates are covered with captured contaminants.

TESTING AND RATING METHODS
FOR PARTICULATE FILTERS

There are a number of filter efficiency testing methods. They include:

- Arrestance Test
- Atmospheric Dustspot Test (ADST)
- DOP Test (IEST)
- Fractional Efficiency (Std. 52.2)

The first two methods were covered in American Society of Heating, Refrigerating Air-Conditioning Engineers, Inc. (ASHRAE) Standard 52.1-1992 which has now been withdrawn by ASHRAE. The Arrestance is the quantity of dust the filter will hold under controlled conditions. It is a gravimetric measurement determined by filter weight gain in comparison to total weight of test dust fed. Thus, it has no relationship to efficiency. This test method is used primarily to differentiate between lower efficiency panel filters. The ADST provides an average efficiency over the life of the filters. It used atmospheric dust as the challenge and discoloration as the gauge of capture efficiency. This test method was widely used for extended media commercial filters.

The DOP test is based on a military specification, which uses Dioctypthalate smoke as the test challenge and a photometer for the determination. Thus, the latter is a fractional efficiency test method and an excellent model. This test method is used to rate ultra high-efficiency filters such as HEPA. Unfortunately, the DOP has been determined to be carcinogenic. Thus, a number of other stearate oils are being used or proposed for an appropriate test aerosol. The cognizant organization for the development and maintenance of test standards dealing with HEPA filters and clean room is Institute of Environmental Sciences and Technologies (IEST).

The test methods promulgated in Standard 52.1 did not address the IAQ needs and are flawed for several reasons such as:

- ADST used an undefined and uncontrolled test aerosol (atmospheric air);

- ADST determination was based on discoloration (a 1930s detection technology). Particle counter technology is now available and much more accurate and sensitive;

- ADST data was presented as an average that distorts the actual filter performance and overstates it for much of the filter life; and

- Arrestance testing yields big and attractive numbers which are confused with ADST data.

The *ASHRAE Standard 52.2-2007: Method of Testing General Ventilation Air-Cleaning Devices for Removal Efficiency by Particle Size*, was developed and promulgated in response to these weaknesses. The test method is based on an ASHRAE- and EPA-sponsored Research Project 671: *Define a Fractional Efficiency Test Method that is Compatible with Particulate Removal Air Cleaners Used in General Ventilation*.

The newer test method is designed to solve much of the confusion and misinformation about the performance of general ventilation filters. The primary characteristics of the 52.2 test method include the following:

- The test method evaluates the *minimum* particle removal efficiency of a clean air-filtering device;

- The test aerosol is based on laboratory-generated KCL particles of a defined uniform size range spanning from 0.3 micron to 10 microns;

- The efficiency of an air-cleaner is determined as a function of particle size. The performance curve is based on 12 size ranges and is determined by an optical particle counter; and

- The test method evaluates *minimum* particle removal efficiency over a full-loading life cycle.

The data are reported using the test method in ASHRAE Standard 52.2 as shown in Figure 8-7, which is a composite curve that incorporates the minimum efficiency at each size fraction bank over the entire multistage loading cycle. This illustration characterizes the minimum efficiency of various typical types of filters according to 12 bands of particle size (0.3 micron to 10 microns). The code on the various graphs is as follows:

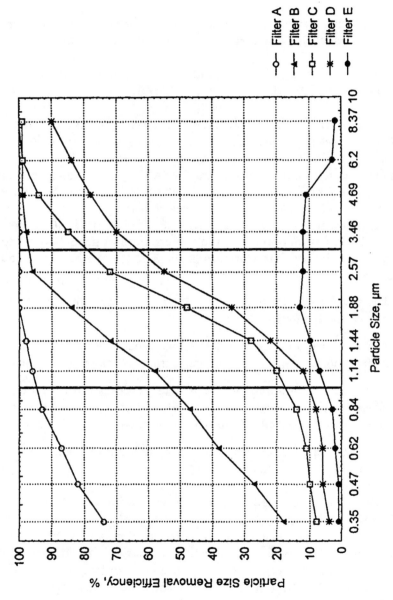

Figure 8-7. Filter Performance Curves of Typical Filters.

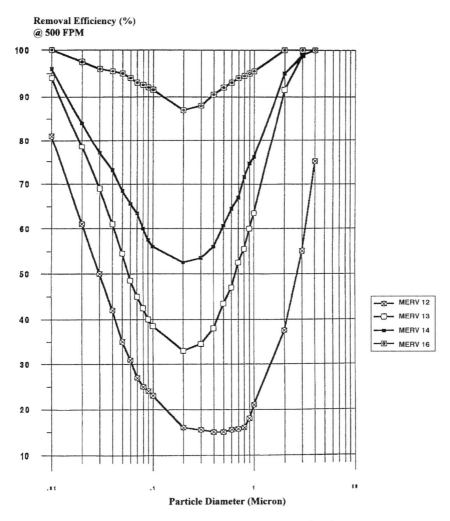

Figure 8-8. Most Penetrating Particle Sizes.

MERV LEVEL	Original Dust Spot %	Typical Particulate Filter Type	% 0.3-1 µm	% 1-3 µm	% 3-10 µm
1	NA	Low efficiency fiberglass and synthetic media disposable panels, cleanable filters, and electrostatic charged media panels	Too low efficiency to be applicable to 52.2 determination		
2	NA				
3	NA				
4	NA				
5	NA	Pleated filters, cartridge/cube filters, and disposable multi-density synthetic link panels.			20-35
6*	NA				36-50
7	25-30%				50-70
8	30-35%				>70
9	40-45%	Enhanced media pleated filters, bag filters of either fiberglass or synthetic media, rigid box filters using lofted or paper media.		>50	>85
10	50-55%			50-65	>85
11	60-65%			65-80	>85
12	70-75%			>80	>90
13	80-85%	Bag filters, rigid box filters, minipleat cartridge filters	>75	>90	>90
14	90-95%		75-85	>90	>90
15	>95%		85-95	>90	>90
16	98%		>95	>95	>95

* MERV 6 level prescribed by Standard 62-2001[19] for minimum protection of HVAC systems

Figure 8-9. Comparison of MERV Data, Filter Type, and Prior Designations.

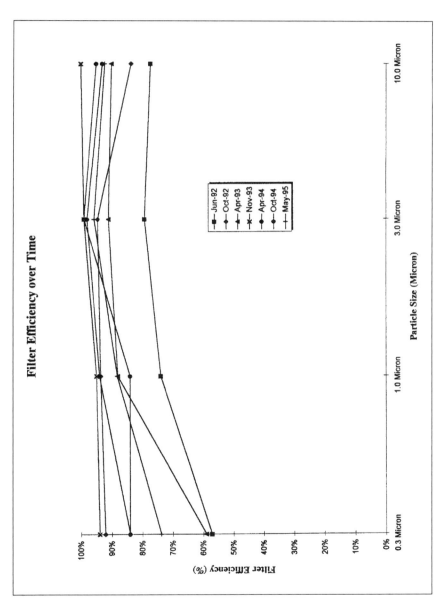

Figure 8-10. Fractional Efficiency of a MERV 14 Mini-pleat Installation Over Time.

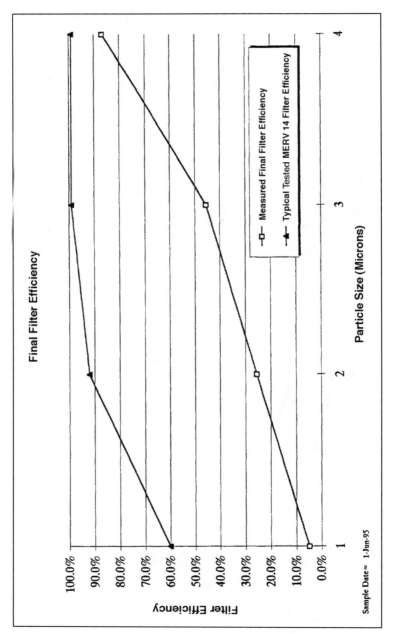

Figure 8-11. Filtration Efficiency of a Typical MERV 14 Filter indicating a filter failure or anomaly in installed performance..

Filter A = 95%	MERV 14
Filter B = 85%	MERV 11
Filter C = 55%	MERV 9
Filter D = 2" pleat	MERV 8
Filter E = Fibrous glass throw-away	MERV 4

The data developed using this test methodology can be helpful in determining specific size penetrations of various media types (refer to Figure 8-8).

The testing method presented in ASHRAE Standard 52.2 is applied under controlled laboratory conditions. However, there is recent evidence that similar techniques and methods may also be applied in the field. Obviously, conditions are less controlled since the challenge aerosol becomes the ambient air. However, the field performance of filtration banks can be evaluated in a similar fashion using the "real world" size fractions of the ambient airstream. Figure 8-10 shows the excellent fractional efficiency of a MERV 14 mini-pleat installation over time. The analysis uses the upstream and downstream particle counting for the determination. This can be compared to Figure 8-11 which typifies the field performance of a pocket filter also rated at MERV 14. Only in this case study, the particle count evaluation revealed that the installed filter bank failed to replicate the expected efficiency of the MERV 14 filter cartridge because several filters experienced severe leakage around their frame and one filter cartridge was only partially installed allowing serious bypass of unfiltered air.

After several years of experience with the new MOT, the SSPC52.2 committee determined that the conditioning step in the original procedure was not sufficiently rigorous to replicate the efficiency drop experienced by certain charged media when exposed to ambient pollution constituents, such as superfine particles and diesel exhaust. Recent research sponsored by SSPC 52.2 has yielded guidance on an enhanced conditioning loading procedure that employs a superfine KCl aerosol as an alternate conditioning step. The revised conditioning method has been published as an Addendum J to Standard 52.2. Thus, designers and owners desiring the more robust conditioning criterion should specify the desired MERV designated with "KCl conditioning" and iterate the specific targeted minimum from the CEC and the testing velocity.

To assist building owners and mangers understand the new MERV designation system, Figure 8-9 correlates the older 52.1 system with the new MERV numbers and describes the generic type and general application of the filters.

Treated Filters

As a part of the furor of concerns over microbial growth in the air pathway, some filter suppliers promoted the treatment of the filtration media with various chemical biostats. The claims were that the additive would retard growth on the filter and potentially improve the quality of the treated air during its life cycle in the air handler. Work by Foarde at RTI, sponsored by ASHRAE, affirmed that such additives did little to alter the growth behavior on filter surfaces in a wet environment. This was due to the cake build-up on the filter media that created a barrier to the biostat and provided a nutrient basis. This feature as been largely abandoned by manufacturers because of the dubious added value.

GAS PHASE FILTRATION

The control technology for gaseous filtration is entirely different than particulate filtration. Particulate filtration is predominantly mechanical and/or electrical whereas gas-phase filtration is chemical. The primary capture device for chemical molecules is sorption. Sorption is a process based on the electron forces carried inherently within the molecule being attracted by similar forces on the surfaces of solid sorption filter beds. These "Van der Waals" forces create powerful bonds by which molecules are attached to the surfaces of a solid matter (adsorption). Any solid surface is a dwelling place for "sorbed" molecules. However, a porous matter provides greater surface area and, thus, more sorption sites. The primary medium for gas-phase capture is a bed of extremely active or very highly porous sorption media, such as an activated carbon. These materials have extremely high internal surfaces that can be measured as high as acres per ounce. This process of attraction and capture onto the surface is defined as adsorption, or clinging "onto" a surface area.

Gaseous control can also be accomplished with absorption or "coming into." In fact, some sorbers such as zeolite and alumina are also very hydrophilic, attracting water at the same time as sorbing contaminant molecules. The combined process of adsorption, which differs from ad-sorption through the process of molecules merging "into" the sorption substrate similar to liquid phase mixtures, while in contact with both water and other reactive molecules, brings about ionization. This enables certain chemical reactions (chemisorption). To enhance this control mechanism, sorbents can be impregnated with specific reagents to enhance their ability to chemically react with gaseous contaminants of concern.

Another approach to attaining chemical reaction is the process of ionization. This employs an electromagnetic field which manipulates the electron charges of airborne molecules enabling them to react with each other. This process is usually employed in conjunction with a sorption bed, and these two processes have been demonstrated to act synergisti-cally. Ozonization is also used for gas-phase control. Ozone is released into the airstream. It is highly unstable and, therefore, a highly reactive compound. Thus, it can chemically modify reactive molecules with which it comes in contact. This process must be used with a sorption bed for the synergy potential and to control the residual airborne ozone. This is important because ozone is itself a regulated contaminant of health concern and should not be used in conditioned space during occupancy.

Control Media Characteristics

The typical control media commercially available include:

- Carbon
- Impregnated carbon
- Alumina/potassium permanganate ($KMnO_4$)
- Zeolite (molecular sieve)
- Zeolite/$KMnO_4$
- Physical blends of the above-listed materials

The factors affecting efficacy of gaseous filters are much more complex than particulate filters, starting with the sorbent media physical properties. Generally, sorbents are in the form of either round cylindrical pellets or random-shaped flakes (refer to Figure 8-12). They can vary in internal surface area (generally the higher the surface area, the higher the capacity). Pellet size can affect apparent surface area as well as airflow and bypass characteristics. The inherent chemical properties of the sorbent can influence chemical contaminant preferences and control capabilities. Another physical property of concern is flammability which

Figure 8-12. Sorption Media with a Sample Bag and Probe. (Courtesy Purafil Inc.)

can be critical when installed in an air conveyance system in a public building.

The contaminants themselves influence the nature and competence of the process. Unlike particulate matter that basically behaves the same (other than size), each gaseous contaminant is unique in its makeup and resulting behavior. The following are some of the factors that influence the behavior of specific chemical gasses in a sorption process:

- Contaminant Mixture
- Concentration
- Molecular weight
- Polarity

- Vapor pressure
- Acidity (pH)
- Reactivity
- Boiling point

Figure 8-13 typifies the variability of sorption performance as a function of varying challenge concentrations.

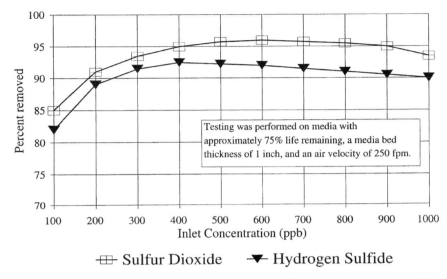

Figure 8-13. Single-Pass Efficiency of KMnO₄/Alumina Against Sulfur Dioxide and Hydrogen Sulfide.

Similarly, atmospheric conditions like temperature, relative humidity (RH) and barometric pressure affect the performance of the sorption bed. Higher temperatures tend to make the molecules more active which tend to overpower the van der Waals forces. For example, carbon is normally reactivated with heat. High humidity can blind the pore structure, and void internal sorption sites by filling them with water molecules. These factors can have significant influence on the behavior of gaseous filters in the field. For example, a 10–°F shift or a 10-point RH swing can influence the sorption capacity isotherm of a carbon filter by as much as 25%.

The containment device for the sorption media will drastically affect the filter performance. This is where many leaps of logic occur in the application of gas-phase filtration. A sorbent is inherently capable of controlling a contaminant; ergo the assumption is incorrectly made that control is true under all conditions of canister configuration (refer to Figure 8-14). Canister characteristics such as media bed depth, air velocity, and the resulting superficial dwell-time within the bed will drastically affect the performance of the filter. Generally, the higher the dwell-time, the better opportunity for complete sorption.

Figure 8-14. Typical Air Cleaner Canisters. (Courtesy Purafil, Inc.)

Evaluation Methods

Rating methods vary depending on the media but generally depend on proprietary data developed by the manufacturer. There is no current concensus industry standard for rating the performance of the filter canister itself. Since each contaminant acts differently with each control media, it is difficult to have a universal test that applies to all conditions. ASHRAE has developed and published ASHRAE Standard 145.1 in 2009 that is a laboratory scale test method to standardize the evaluation and/or screening of various pelletized sorption media. A second phase of the Standard 145.2 addresses full scale filter canister devices using an adaptation of the ASHRAE 52.2 test duct. The latter Standard has been approved for publication in early 2011. Current rating methods include:

- Carbon Tet No.
- Isotherms
- Single point evaluations
- Break-through curves

- Single contaminant curve
- ASHRAE Standard 145.1-2009
- Physical properties
- ASHRAE Standard 145.2-2011

Since there are no universal test methods for rating and selecting sorbent product, and the Standard 145.2 has little established history, the following discussion may be helpful in the selection process.

APPLICATION OF GAS PHASE/ CHEMICAL FILTRATION

Carbon is traditionally considered the "universal sorbent" because of its long history going back before WWI gas mask applications along with its high surface area—as much as 2500 M^2 per gram of carbon media. The general control properties of this type of media that are described below can be enhanced or "treated" with reagents that will modify the ability of carbon to control specific contaminants of concern, such as war gasses or certain toxic industrial chemicals. It is applied best against contaminants that are higher in concentration, heavier in molecular weight, nonpolar, and having a high molecular-weight of carbon molecule content. This includes most volatile organic compounds (VOCs) and long molecular-chain hydrocarbons. It also prefers lower atmospheric temperature and humidity. The $KMnO_4$-modified substrates perform best against lower molecular-weight compounds, polar-compounds such as formaldehyde, and reactive inorganic compounds such as acid gases like hydrogen sulfide (H_2S) and sulfur dioxide (SO_2). Zeolite is particularly effective as a cation exchange media. This makes it perform well against contaminants like ammonia and other nitrogen-bearing compounds. It also has higher surface area than alumina compounds, but is somewhat softer and more prone to off-dusting.

These materials are all in pellet form, varying in size and shape. The application in the air stream can vary dependent upon required efficiency, pressure loss constraints, and life cycle expectations. When applied as slurry coated foams or fabrics, the filter is lower cost and lower pressure drop. However, the efficiency and performance is very low and short-lived with breakthrough and rapid decrease in efficiency occurring within minutes of contaminant exposure. This is because of the limited amount of active sorbent actually applied in the matte. Improved performance has been demonstrated when the sorption pellets are loaded or "immobilized" into non-woven fabric that can then be pleated into extended media filter cartridges. These medium efficiency filter cartridges can be applied to light duty commercial applications. Although they demonstrate an early breakthrough and fractional efficiency thereafter, their active sorbent content is higher than slurry coat versions, which yields increased capacity. For the more demanding applications, loose fill canisters are employed in various solid bed thickness up to three inches. To lower pressure loss and increase dwell time, these beds are corrugated or pleated into 12", 18" or 24" deep retainer frames. In even higher concentration or demanding application, solid bed scrubbers can employ the loose fill pellets in even deeper "pour-in/pour-out" beds. In any of the deeper beds, the pelletized media types can even be co-blended to attain a merger of control properties. The deeper beds will generally provide total retention (i.e. no breakthrough of contaminants of concern) and long life cycle because of the mass of sorbent applied to the air stream. The trend of these various filter canister alternatives is—that cost, size, and pressure drop increase as the depth and complexity of the filter bed increases.

Figure 8-15 shows a typical side access air-cleaner unit equipped with pre and final particulate filter banks in addition to the sorption section.

Figure 8-15. A Typical Side Access Air-Cleaner Unit. (Courtesy Purafil, Inc.)

SELECTION AND APPLICATION OF
FILTRATION AND AIR CLEANING

The following decision paradigm will provide the designer with an effective process for selecting, specifying, and applying filtration and air cleaning equipment. An evaluation of the building site and an understanding of the intent and content of the building will assist in predetermining known or predicted contaminants of concern.

• *Identify contaminant(s)*—An evaluation of the building site and an understanding of the intent and content of the building will assist in predetermining known or predicted contaminants of concern. Usually the air is a complex gas that includes numerous contaminants. The nature and content of the complete airstream must be known to develop an appropriate system and select proper filtering media. This should include the contaminant nature, concentration, toxicity and health risk. Also, other atmospheric properties of the airstream must be understood to aid in the proper selection. An evaluation of outdoor environmental pollutants may be required to comply with Standard 62.

- *Establish control end point objectives*—Select control targets, such as: no perceptible odor, no more than 10,000 particles per cubic foot of respirable size, or no greater than 0.1 parts per million (ppm) H_2S for corrosion protection of electronic gear. This would also include an assessment of the applicable standards and the mandatory requirements that they may impose upon building control targets.

- *Select control strategy*—Start with the selection of a filtering device or system. This includes the type, media and efficiency expectations.

- *Perform Life Cycle Cost Analysis (LCCA)*—Before selecting a specific system or device, perform an LCCA. The initial selection may perform adequately, but LCCA may reveal short service life, resulting in higher maintenance cost for the owner.

- *Locate it in the system*—Dependent on the contaminant source, type of mechanical system, site of exposure and control strategy.

- *Specify and select competent seal*—This is the common weakness in any filtration system as a significant percentage of air can by-pass the filter cartridge, especially in higher pressure drop versions.

- *Size for value*—By over-sizing the filter bank, the operating cost can be substantially improved because of disproportionate gains in life cycle, dirt holding capacity and operating energy.

- *Select the proper cartridge or canister*—Even when specific efficiency performance is determined, varying filter configurations can impact overall performance including life cycle, pressure drop, and LLCA.

- *Establish airflow path*—This involves how the air will be delivered to and from the filtering device, how much air will be treated, and how it will be delivered to the targeted zone.

- *Build in space and capacity for retrofit*—This allows for what may happen in the future. Perhaps the ultimate activity is unknown, the number of tenants are undetermined, or future expansions.

- *Evaluate use of prefiltration*—According to conventional wisdom extended media particulate filters should be prefiltered. This is based upon the generally held belief that prefiltration protects the final filter, extending its useful life proportionately. Anecdotal experience by this author indicates that the additional cost, room, labor, and energy consumption do not cost-justify the marginal gain of filter life. However, it is well advised to protect by prefiltration the following filter types: HEPA, gas-phase sorbers, minipleats and EACs. The use of prefilters on the other filter types can increase cost, labor, energy, space, and add little in extended life or value.

- *Install adequate filtration during construction*—The filters of design level should be installed when the contamination and/or exposure levels occur at their highest peaks. Thus, the filtration system should be installed during the construction period when a major portion of the building distribution system contamination occurs. This is contrary of most contractor practice because they often use inexpensive disposable filters during the construction period when the contamination load is the highest in the life cycle of the building. Then, the specified filters are installed at turn-over with the result being that the distribution ductwork and HVAC system are pre-contaminated with construction particles.

- *Provide for monitoring performance of filtration system*—This can be done in situ with monitoring equipment that can signal failure or service needs to the building management control system. Periodic evaluation of the space will best ensure that a high level of air quality is sustained in the space.

- *Consider building security issues*—Filtration and gas phase air cleaning can play a significant role in the strengthening of building security and protection against either accidental or criminal exposure to airborne contaminants. Because of the importance of this subject to building owners and the close relationship of this emerging need with IAQ, a special chapter on this subject has been added to this edition of MIAQ. This discussion will include discussion of the role of filtration and provide specific references for further study.

- *Consider the IAQP*—once the decision is made to invest in enhanced filtration and/or air cleaning for any reason, the IAQ Procedure of Standard 62 should be evaluated. Code variance may be required, but the energy savings from lowering the heat load of outdoor ventilation air provides compelling motivation. Even if the HVAC system is sized and built to code required capacity for prescribed ventilation levels, the installation of enhanced filtration will enable the on-going operation of the facility under the IAQ Procedure with reduced ventilation. Especially consider the IAQ Procedure if there is poor outdoor air quality; outdoor air is frequently hot and humid; the facility has a high occupant density or experiences wide fluctuations and diversity in space usage; when existing refrigeration/ air handling capacity is limited; and when the Contaminants of Concern are known or are easily ascertained.

The following chart Figure 8-16 is adapted with permission from the ASHRAE IAQ Design Guide and summarizes the selection of filtration and air cleaning for specific design strategies or situation. These general recommendations are based on efficiency, thus, the reader is cautioned that the final selection must also include considerations of the physical and mechanical limitations of the air handling equipment; the characteristics of the Contaminants of Concern; and the overall factors that concern LLCA, such as system life and pressure drop.

ACKNOWLEDGMENTS

The revision author gratefully acknowledges the valuable contributions of the following entities for this chapter: The Filtration Group, Inc., Joliet, IL, for their contribution of proprietary laboratory test data and generic product photographs; Environmental Design International, Ltd., Atlanta, GA, for proprietary field test data and critical contributions in the illustrations; Purafil, Inc., Atlanta, GA, for their contributions of generic gas-cleaning photographs; National Air Filtration Association (NAFA), for permission to reproduce essential drawings and charts; and ASHRAE, Atlanta, GA, for their permission to reproduce charts.

Figure 8-16. General Selection Guidance Based on Efficiency

Design Strategy	Recommendation/ Solution Range	Outcome/Benefit
Keep heat exchange and distribution system surfaces cleaner	Apply MERV 8–13	Reduces excessive coil fouling; maintains heat exchange efficiency; reduces energy losses; lowers cleaning and maintenance costs; maintains system performance; lessens opportunities for mold growth by reducing nutrient and moisture retention
Keep conditioned space cleaner	Apply MERV 8–13	Reduces nuisance dust; lowers housekeeping costs
Control viable and/or pathogenic particles	Apply MERV 14–16	Lowers occupant exposure to airborne pathogens; decreases negative health effects; decreases absenteeism (Milam 1992) and related productivity costs
Treat excessively polluted outdoor air	Apply MERV 11–14 Apply Gas Phase ME	Ensures acceptability when outdoor air is non-compliant; lowers occupant exposure to external sources of particles, odors, and irritants; lowers risk of cross-contamination from building exhausts; reduces undesirable products of chemical reaction between ozone and indoor chemicals
Control specific CoC, including those associated with criminal intent or accident	Apply MERV 13–16 Apply HE-HEGA	Reduces risk of accidental spills or criminal incidents of particulate or gas-phase contamination that are detrimental to processes, products, people, or their related activities
Augment outdoor air ventilation rates using the IAQ Procedure from ASHRAE Standard 62.1 (ASHRAE 2007b)	Apply MERV 11–13 Apply ME-HE	Can reduce excessive latent load from outdoor air in certain regions; reduces contaminant load from polluted outdoor air; can result in reduced HVAC capacity and related capital cost, energy consumption, and operating cost

Notes:

MERV = Minimum Efficiency Reporting Value

HEPA = high-efficiency particulate air

ME = medium-efficiency range partial bypass gas-phase air cleaner

HE = high-efficiency solid bed gas-phase air cleaner

HEGA = high-efficiency gas deep bed adsorber

Chapter 9

BUILDING SECURITY AND SAFETY
Readying Your Building for "Extraordinary" Environmental Stress

FOREWORD

This chapter has been included in *Managing Indoor Air Quality* because of the close relationship of building environmental security and safety to the core issues of IEQ. This discussion also recognizes the need for practical and cost effective counsel for building owners and managers to deal with the growing risk of exceptional and extraordinary environmental challenges—whether they be from accidental or criminal causes. The initial panic and knee-jerk reactions to Homeland Security issues have now tempered, but the lessons learned from the 9/11 bombings and subsequent Anthrax attacks were costly. The material in this chapter is intended as a management overview and cannot, therefore, include full details about all of the potential threats, their health effects, and all of the prevention or remediation options available to building management. However, as you will note in the following discussion, there are significant parallels in evaluation, prevention, and operation in dealing with both IAQ as well as building environmental security issues. Because this book is focused primarily on the indoor environment with further focus on the air pathway, issues of building security that have to do with infrastructure, such as food, water, and electricity are not covered. In a related fashion, issues like blast-proofing, building accessibility, and occupant entry and egress are not included. As forward thinking managers, however, it is important to include all of these issues in your budgeting and operation planning. As an aid to the reader, there are specific references on this subject located at the end of the chapter.

"9/11" CHANGED THE WORLD

It has been said eloquently, emotionally, and frequently that the world is now different for building owners and managers because of the events of 9/11/01. In many ways this is true. Yet, 9/11 did not change the size or nature of a 1 micrometer-sized particle or how it behaves. It did not change molecules and make them behave any differently. It did not alter the physics of air movement and treatment. Nor did it change how a building responds or does not respond to internal or external challenges to its performance or ability to provide an acceptable and healthy indoor environment for its occupants.

The early reaction to the events surrounding 9/11 was similar to the "Chicken Little" of nursery rhyme folklore fame. Again the shouts of "The sky is falling!" spread panic and concern among the building community. Conferences were held by every conceivable trade group, publication, and government agency related to the building industry; legislation was proposed at Federal and city level; a special government "super" agency was created; a national alert system was developed; and duct-tape sales surged. In 1974, the label for a similar panic was "Energy Conservation" that brought about the thermostat police. In the 1980s, the label was "Tight Building Syndrome" and "IAQ." In the current decade, today's label is "Anthrax Spores" and "CBR" exposure (Chemical/Biological/Radiation). These are indeed valid reasons for concern as they can be life threatening. The fear has been kept alive by the SARS outbreak in China in 2003 that demonstrates how far and how fast a communicable disease can spread in our modern world made small by a mobile society. However, the response to these issues must not be panic reaction, but rather, rational planning and preparation. The purpose of this chapter is to help the building management team respond to the challenge of this new "piece of falling sky" issue in a rational, reasoned and proactive manner.

The Issue is Complex

Building management has never faced so immense, so complex, and so critical an issue as environmental security. The reason that it is so *immense* is that it involves the external environment with its implicit environmental, infrastructure, and societal issues; the entire building stock; the individual building itself with all its structural and operating systems; the occupants with their operations, policies, and activities; and the building operation and maintenance practices. It is *complex* because there are so many faces to the challenge; there are so many avenues through which it can enter and/or challenge our buildings; and there is such a wide range of alternative responses that could or should be imposed upon the building and its operating systems. It is *critical* because as stated by ASHRAE Past-President William Coad (2001/2), "It raises the issue of the quality of the indoor environment from a comfort and housekeeping issue to a health and safety issue." It is also *critical* because it affects our basic culture and way of life that is based upon a free and open society with easy and open access to our public buildings.

In order to bring some perspective to this matter, however, building managers must consider that the incidence of an exceptional environmental challenge is a relatively rare occurrence. Whether the criminal activity of chemical or biological release, or the accidental spill of a tank-car of industrial chemicals at the next street corner, these events are unusual when considered in light of the millions of commercial buildings and relative infrequency of their occurrence. However, according to the report of ASHRAE Presidential Ad Hoc Committee on Building Security, the incidents have been increasing dramatically on a world-wide basis over the last several decades. Thus, the consideration of the hardening of your building becomes an essential matter of risk assessment, exposure, and risk management. The building's public profile and image, location, size, age, historical significance, and adjacent potential exposure sources are all factors that will influence your particular analysis of economic risk. Similarly, it will drive your response tactics in adapting your building to deal with the determined risk. To aid the building owners to assess their risk and to apply known tactics to help minimize their vulnerability, the work of the Ad Hoc committee was incorporated into a published document entitled, ASHRAE Guideline 29-2009 Guideline for the Risk Management of Public Health and Safety in Buildings.

The Rules Have Not Changed

This guidance is written from the perspective of the building manager who must deal with the everyday demands of operating facilities and HVAC systems; the constraints of system limitations and capacity; and the limitations and realities of budgetary and energy costs. To set the stage and offer perspective on the issue, it is important to consider some basic guidelines and principles in your risk assessment and evaluation process.

Do No Harm

As other professions dealing with the health and welfare of the general public, the building managers and engineers must follow this general guidance as the first rule of reactionary response to exceptional challenges—do nothing that will result in further harm. In 1974, the cry was to "Seal off outdoor air ventilation systems!" in order to save energy. That prompted and created the negative reaction of the first rash of IAQ problems resulting from TBS (Tight Building Syndrome, an early label for IAQ problems from ventilation-deprived buildings). That same cry came from New York City as an outcome of the initial panic from the 9/11 attacks—and the result would have been more "harm."

The Building is a Holistic Entity Consisting of a
Complex Matrix of Systems

Like the human body, building performance is based upon interconnected systems and interrelated functions. When one component is "ill" or broken, the whole building performance suffers. When the legs are not well, the organism cannot run. When the building control system has a "headache," the environmental control system fails. It is this very matrix that makes the building work well that makes it vulnerable to indiscriminate "tweeking" that can harm the entire organism.

For Every Action,
There is an Equal and Opposite Reaction

This is a classic cause and effect situation. Quick fixes, like Band-Aids, can cover up a problem and conceal the real outcome—but not for long and with potential harmful effect somewhere else. For example, raising chilled water temperature to save energy can result in the loss of control of humidity resulting in mold infestation. Reducing outdoor ventilation air to reduce energy cost or to reduce outdoor air exposure potential can allow internal contaminants to accumulate and create discomfort, IAQ complaints, and potentially adverse health effects.

Air is a Non-compressible Fluid

In the context of a building, air can be pumped into or out of a building, but, generally speaking, it is neither compressed nor is a vacuum created. Thus, air follows small incremental pressure differences to create the air pathway. As we will see later, this is a critical concept as the air pathway is very critical to sustained indoor air quality conditions.

Air Follows the Negative Pressure

The more powerful of the drivers in the air pathway from a direction and control aspect is the negative part of the equation. Thus, the return and/or exhaust systems of a building become the most dominant controller of the air pathway, and thus control the pathway of unwanted contaminants.

A 1 Micron Particle is a 1 Micron Particle

Whether particles in the outdoor air are Anthrax spores or condensation nuclei from automobile exhaust or tobacco smoke, they still behave the same and are potentially respirable into the lungs of the occupants. They both represent health risks to the occupant. The one may be viable and pathogenic, the other may be carcinogenic. However, they behave the same when responding to control tactics like dilution and, especially, filtration. Outdoor air contaminated with respirable particulate matter of less than 2.5 microns in size is a serious challenge to the indoor environmental quality. This was true before 9/11 and it remains true today.

Assume Nothing

When you break the word "assume" apart into three syllables, the result (making an _ _ _ of you and me) will never be truer than now. The Building Management System computer can, does, and will provide inaccurate information. Don't assume anything that has not been investigated by personal "line of sight" techniques and a thorough and detailed "been there, done that" walk-through of the building.

One Size Does Not "Fit All"

Simple universal fixes are seldom possible because of the wide variations of building architectures, envelopes, HVAC systems, occupancies and activities, building usage, and operating/maintenance characteristics and practices. Thus, every building must be evaluated "where-is: as-is" to assess the risk and understand the needs and potential for protection.

Good CB Preparation is Also Good IAQ

Generally, those things that are performed for CB protection will also bring about improved system cleanliness, improved indoor air quality, and improved system performance. For example, good filtration creates and sustains cleaner coils and systems—systems that perform well from an energy standpoint—that resist microbial growth from an IAQ standpoint—and that decrease concentrations of airborne bacteria.

ASSESSMENT GUIDANCE

The following elements provide a brief guide to building management to assist in assessing the needs of their facility to develop a risk assessment and to prepare for unusual environmental challenges. These suggestions are based upon those actions that can be taken immediately and focus on current systems and operational practices. The resulting recommendations presume minimal capital expenditure or structural invasion. The process outlined will also help define areas that would provide good value-added features with more extensive or elaborate capital and engineering support.

Understand How the Building Works

This is probably the most basic and important baseline information that you can develop. Even if you think you know everything about the building and how it works, you have to know it so well that you can explain it to a bureaucrat from Homeland Security who is telling you to do something irrational that will create adverse consequences on building operation. *(Remember the cause and effect rule)*

Get Help from a Professional

Even if you have a "history" with the building, it helps to bring in an expert who will spot "warts" in your building that your facility management team has accepted over the years. This is especially important if you are new to the building, lack original drawings or specifications, have no recent TAB data, or lack an effective Building Management System. Your outside professional should understand the "Rules" cited above and especially understand issues of air pathway, pressurization, filtration, and system response. *(Remember, do no harm.)*

Do Nothing to the Building in Response to Panic

Fear and panic can easily override concerned rational response and will create bad regulations, bad laws, bad policies, and bad instructions (Remember Homeland Security and the duct-tape issue). The purpose of this checklist is to help you build a rational and effective deterrent program and to assist you in defending your position with investors, occupants, tenants, or regulators. *(Remember cause and effect and do no harm.)*

Ensure that the Building is Operating as Intended (or Altered)

Easier said than done, this knowledge is basic to your understanding of the current building performance and to its performance under exceptional challenges. If "that thing has never worked right" is the outcome of this review, then the system response to severe changes of operation is unknown and unpredictable. Are alterations of systems, space usages, or occupancies documented? If there are more questions instead of answers to system performance levels, then recommissioning or new TAB service (test, adjust, and balance) may be required to establish a new baseline of performance. For example, if the roof leaks water, and the envelope leaks air, why worry about exceptional environmental exposures until these basic elements are repaired and operating as they should. *(Remember, holistic entity and assumptions)*

Review the Occupant Egress Pathway

Walk the pathway out of the building. Is it clear, accessible, and well ventilated? If ventilated by the smoke evacuation system, where does the air come from and where does it exhaust to? How is the evacuation system activated and deactivated? How does this system interface with other mechanical systems that may be operating concurrently?

Review the "Deferred Maintenance" List

This list usually consists of those things that really ought to have already been done, but lacked funding in prior fiscal periods. It may include in the listing some things that must be done in order to equip the building to properly respond to those changes required for CB response and control. Thus, these actions must be taken to bring the building back into predictable performance.

Focus upon the Air Pathway

Investigate, track, and understand the air pathway from entry into the building—to distribution throughout the conditioned space—to exhaust from the building. This is critical as it is the exposure route to occupants and/or is the elimination and control device for any unwanted contaminant within the building. The following items are subsets of the air pathway-involved systems. *(Remember, follow the CFM)*

Air Pathway: Focus upon the Air Source—Understand the external factors of the building site as it pertains to the air pathway. Outdoor air introduced at street or subterranean level is the most physically accessible to outside intervention, as well as the most polluted source of outdoor air. Roof air entry is the least accessible and a cleaner source. Are there setbacks or barriers to prevent pedestrian access to the building and the air pathway? Is the air entry located at a dock area that is easily accessed by numerous outside vehicles? Is the entry point near other pollution or fan-powered air sources, such as cooling towers that could be contaminant vectors?

Air Pathway: Focus upon the Air Entry—Regardless of location, what kind of outdoor louvers are in place? Is the air pathway clear? Are the louvers fixed or movable? Will they move properly or have they seized-up? Will they fully close and will they seal when closed? If not, how much leakage is likely with the air handlers operating? What drives the status of louvers if they are not fixed? Are they manually altered? Are they controlled by occupancy or demand monitors? Is an economizer cycle in place? Or is make-up driven by exhaust? These are issues that must be fully understood when a building technician is instructed to "Simply shut off the outdoor air system." As part of this evaluation, understand the role of infiltration as it influences your specific building. Measure or calculate the best you can the factor contributing to the net infiltration volume and pressure, such as wind velocities, glazing condition, envelope penetrations, seasonality, and stack effect. The determination of infiltration is largely an "art form" as the influences are highly variable and highly dependent upon a wide range of velocity pressure dynamics that are in constant flux and are different for every building.

Air Pathway: Focus upon the Filtration—One of the most important possible deterrents to airborne contaminants is the filtration system. Although, the type of filter and its efficiency is important, of equal importance are the retainer system, seal, and bypass. Many systems now employ MERV 5 to MERV 6 Pleats. (Minimum Efficiency Reporting Value is the data product of the filter efficiency test method ASHRAE 52.2.) This MERV 6 filter is approximately 40-50% efficient against 3 micron-sized particles. Higher efficiency pleated filters in the MERV 8 to MERV 11 range will increase efficiencies to as high as 60% against 1 micron sized particles (the lower size range of Anthrax spores). However, if the filter is retained in an individual retainer frame, the seal gasket often takes a permanent set or comes loose or even crumbles with age. Incorrect or inoperative retainer clips allow filter cartridges to fall out or by-pass air. Side access frames don't seal well between filters and often allow serious gaps for by-pass at the access doors. These flaws can impair the overall performance of the filter bank regardless of the rated efficiency of the filter cartridge. Upgrades to higher efficiency particulate filters or gas phase air cleaners are possible but require careful analysis of adequate room, impact on airflow, and sufficient fan-power capacity to overcome pressure drop. More details on filtration are included below and in the filtration chapter.

Air Pathway: Focus upon the Air Handler—If the air handl`er is a unitary self-contained unit like a roof-top unit, where is it located and how accessible is it to unauthorized personnel? What are the entry and exhaust pathways? If it is located in a mechanical room, is the return air pathway through the MR (mechanical room)? How accessible is the MR to unauthorized personnel? Is the MR clean and clear of rubble and extraneous stored material that might impair immediate access to the controls or mechanical components or impair cleaning should an exposure occur?

Air Pathway: Focus upon the Air Distribution—Knowing what air handler serves what zone will go a long way in understanding the potential pathways and exposure potential of exceptional environmental challenges. What air handler serves the mailroom and is it on a separate zone? What is the return air pathway from critical zones of elevated risk? Where are the high population gathering areas, such as the cafeteria and the auditorium and what are their related air pathways? What are the general pressure relationships between zones and the air pathway? What aberrant pathways are possible between zones, such as elevator shafts or stairwells? What are the current ventilation requirements for various zones per Standard 62 and are they in compliance?

Air Pathway: Focus upon the Air Exhaust System—The exhaust system will often dominate the amount of outdoor air required for make-up (and not the ventilation volume required by Std. 62). It is also the exhaust system of the building that will over-ride any attempt to "close down the outdoor air system." So long as any portion of the exhaust system is operative, a comparable volume of outdoor air will enter the building via infiltration through whatever opening it can find *(Remember air is non-compressible and follows the negative pressure)*. This route will be through the air entry louvers if they are not sealed, and/ or open external doors, and/or windows, and/or cracks in the envelope. It will carry whatever integral heat load and airborne constituents it bears into the conditioned space without treatment, conditioning, or filtration. Thus, it is extremely important to understand completely the actual sizing and operation of all exhaust systems. It is important also to realize that high rise buildings have a built-in exhaust system called stack effect—which is powerful enough on a day of extreme temperature differential to over-ride even the exhaust system and positive pressurization system. Further, it is important to be cognizant of the eventual pathway of the exhaust air stream—to assure that it does not re-entrain back into the building by ingestion or infiltration or be a potential hazard to occupants fleeing from the egress route.

CHECKLIST FOR CB HARDENING PREPARATIONS

The following checklist is a building hardening/response "hit-list." The listing generally summarizes the collective recommendations of experts, such as ASHRAE, BOMA, US Army Eng. Corps, NIOSH/ CDC, AIA, and IFMA. Time and space do not allow for detailing all related recommendations but this guidance will assist in the initial assessment and response.

Outdoor Ventilation Air

Do not shut down the outside air systems—without understanding the related systemic, building performance, and contaminant control implications of such action. For example, if the contaminant source is generated in the interior, dilution has been determined to aid contaminant reduction. Thus, outdoor air supply should not be reduced, as it is the removal control mechanism. If outdoor contamination is the suspected source, then shut-down is indicated, so long as related systems don't cause uncontrolled infiltration from other sources.

Outdoor Air Entry

Make ground entry air pathways as inaccessible as possible to contaminant sources—by restricting access, securing, heightening louvers, concealing, relocating, securing, or other means of protecting the air entry pathway from potential criminal or accidental harm—without restricting normal airflows.

Determine Likely Sources and Pathway

Survey and determine likely or potential exceptional sources—from both external and internal sources, personnel and air entry, building distribution pathways, and internal activities.

Access to the Building

Apply stringent clearance and security control methods on accessing the building, especially critical areas such as mechanical rooms, roof, and dock areas. Limit traffic from the public areas to the internal zones of the enterprise. The dock is vulnerable since it is the routine portal of public commerce and routine deliveries—from the postman—to the pizza guy—to UPS.

Building Pressurization

Attain and sustain proper pressurization relationships within the building—positive to atmosphere, properly balanced with exhaust, and balanced properly between zones.

Life/Safety Systems

Verify the operation and appropriate response under duress of all life/safety equipment and related control and response systems—including fire detection, fire suppression, smoke evaluation, and related control devices. An actual occurrence is the wrong time to discover that fans are running backward; control sequences are incorrect; or doorways to the stairwell are locked.

Air Pathway

Ensure that the entire air pathway is clear of debris, obstructions, and flammable material—including mechanical rooms and exhaust systems. Assure that all related mechanical equipment is operative and responsive. Inspect and assure that outdoor air entry is protected or isolated from external access.

Controls and Operation

Verify where and which controls do what in mechanical sequences and fan operation protocols—correct weaknesses as required.

Air Filtration

Upgrade particulate air filtration to as high MERV efficiency numbers as possible—considering retainer/tracking, available space, access configuration and available fan capacity to deal with increased pressure drop. (See later discussion in this chapter and the filter chapter more information on filtration and air cleaners.)

Emergency Occupant Egress

Select and verify egress pathways—with appropriate emergency ventilation with contaminant-free air; and related occupant authority, instructions, communication, and *training*, and *training*, and *training*.

Safe Haven Areas

Such sanctuaries are possible but difficult and costly to designate, isolate, design, and install in most ordinary commercial buildings—particularly for large populations. The safe haven may, in fact, be contrary to the appropriate safety measure, which may be to vacate the building. Emergency readiness for such spaces requires abnormally rigorous and sustained maintenance of life support and environmental protection equipment that is not normally available or applied in most buildings. Special purpose and ultra-critical spaces can be protected with specialized safe-haven techniques—HEPA and gas phase filtration, pressurization barriers, isolation and sealing, special access and egress, emergency power, and related life support survival supplies—all of which must be sustained in readiness on a 24/7 basis. Such tactics would be outside the normal operating budgets and maintenance capabilities of commercial buildings.

Monitoring Activity

Though not currently available for commercial application because of response time, sensitivity, and cost, building managers should stay alert to new developments in the monitoring field. Pressurization monitors are now available to provide alerts for air barrier failure. As the technology advances under the sponsorship of Homeland Security grants, building owners should apply contaminant monitors to their system response control regime.

Preparedness Plan

The above checklist elements should be thoroughly reviewed, resolved, and written for your facility—including designated authority(s); response protocols, policies, and practices; instructions and training, and building systems response tactics. Management should thoroughly and repeatedly communicate these plans to occupants and tenants.

The brief list above substantially understates the issue of hardening a building because such a "Hit-List" is easy to write—much harder for the facility management team to do. Each of the items involves line-of-sight inspections both before and after. It takes a full understanding of design intent and original plans and specifications and many times these do not exist. It requires "in situ" evaluation and demonstration of systems and components in full and part load conditions—both qualitative and quantitative. It demands an understanding of the rules of physics and air behavior described earlier in the chapter. It may require outside professional studies, including potential demonstrations to verify system correction, alterations, or upgrades. Further, it requires of management the creative input and support to perform risk assessment decisions and to support renovation budgets and operational upgrades. Hardest of all, it involves change... changes in attitude by both management and operating technicians... changes in priorities... and changes in importance of the role of the indoor environment in the view of building management and long term security.

FILTRATION IS AN IMPORTANT
BUILDING SECURITY TOOL

An area of HVAC technology that has long been under-utilized in commercial buildings is the area of air filtration. Relatively few commercial buildings have high efficiency filtration equipment installed and even fewer buildings have molecular/gas phase filters installed. The recent criminal events of airborne biological exposure in public buildings have heightened the interest in improving filtration because high efficiency filtration and air cleaning is one of the few proactive techniques for controlling airborne weapons of mass destruction. However, like the other issues in the above "Hit-List," filtration upgrades should be driven by thorough risk assessment analysis and related cost/benefit analysis.

Assuming that filtration improvement is supported by risk analysis, the following recommendations will aid in the selection and upgrading process. Although not an over-riding "silver bullet," enhanced filtration and air cleaning can add greatly to the building security capability. General information about filters is covered in the chapter devoted to this technology. However, the following comments are pertinent specifically to building security and are included or reiterated for this discussion.

Understand the Air Pathway

Prior to filter enhancement selection, understand fully the air handling system capabilities and capacity; filtration efficiency requirements; and the nature, source and probable pathway of the contaminant(s) of concern.

Evaluate the Filter Retainer System

Prior to any filter modification or upgrades, physically examine existing filter installation banks, including slide track, retainers and access doors to ensure proper fit, seal, and avoidance of by-pass. Experience indicates from 5% to 25% air by-pass and leakage around the filter retainer/tracking system in typical commercial buildings. By-pass around or between retainer frames, around filter cartridges, or between filters and access panels must be repaired and sealed in order to gain the control benefit of the extraction process.

Seal the Filter Cartridge

Gasketing must be examined for integrity and resiliency, and flawed seal surfaces repaired and/or upgraded to resilient gasketing material, such as neoprene.

Seal Side-load Units

Slide-in filter units should be gasketed and cartridges sealed between mating filter frame surfaces to avoid air by-pass between modules due to frame distortion and dimensional variance.

Upgrade to Higher MERV Filters

Dependent upon the constraints of retainer system, sizing and available room, airflow capacity, and pressure loss considerations, particulate filtration should be upgraded to the highest MERV designation that can be practically and physically accomplished without system modifications. (See the related discussion below and in the filtration chapter.)

Upgrade 2" Pleats to Higher Efficiency Media

Most 2" pleated filters that demonstrate enhanced MERV 8-11 designations can operate within the existing AHU constraints of airflow, size, and pressure drop. Though more expensive than lower performing MERV selections, these upgrades are feasible with relatively low additional operating cost premiums. Thus, they can provide substantial improvement in particle control without system modification and represent a cost-effective way of enhancing air quality and cleanliness.

Upgrade 4" Pleats to Higher Efficiency Media

When front-loading retainer systems or 4" side-loading tracts are available, upgrade to the highest MERV efficiency 4" pleat for a cost-effective upgrade without system modification. The deeper cartridge provides up to double the surface area over 2" deep versions, which enhances both efficiency and life cycle. Substantial operating savings are also accrued due to decreased pressure drop, longer life cycle, and sustained cleanliness of the HVAC system.

Upgrade to Highest Efficiency Media Possible

When front-loading retainer systems or 4" side-loading tracts are available, and higher efficiency is desired, consider upgrading to 4" deep mini-pleat MERV 13 or MERV 14 cartridges. Although significantly higher first cost and higher pressure loss than the lower efficiency conventional pleats, these filters will provide higher efficiencies without physical system modification, although airflow restrictions may be encountered.

Upgrade Current Bag/Cartridges Filters to HEPA Range

If current filtration systems consist of extended media bag or cartridge filters in the MERV 13 or 14 level, upgrade to MERV 16, which is an efficiency approaching low-end HEPA filtration (95% against .3

micron sized particles). The MERV 16 filter is available in the mini-pleat configuration that will provide substantial improvement in efficiency while operating in the same pressure drop range as the lower efficiency filters it replaces.

Treat Outdoor Air Ventilation Systems Separately

If separate outdoor ventilation systems or separate outdoor air filter banks are installed, the related filter banks could be enhanced independent of the overall air handling system. Higher MERV designated filters should be employed selectively for outdoor air treatment, which would minimize overall filter cost and isolate/limit any system or filter bank modifications.

Filter Specialty Spaces

Especially engineered or modified systems could provide air delivery selectively to selected spaces, such as egress, sanctuary locations, or mail rooms. This would incorporate very high efficiency filtration and molecular/gas phase air cleaning equipment. Similarly, engineered self-contained filtration units could be positioned in a by-pass mode to treat specific airstreams on demand, such as outdoor air. Such systems could operate independent of the central HVAC system.

Use Filter Monitoring to Drive Change Cycles

When extended media filters are installed (MERV 6 and higher), air filter gages should be employed to monitor the pressure drop performance of the filter cartridge. This is the most cost-effective way to assure the correct life/change cycle for servicing filtration systems. Visual inspection or arbitrary timed change cycle routinely shortens the potential efficacy and life cycle of filter cartridges.

Use LCCA (Life Cycle Cost Analysis) in Selecting Filter Alternatives

Examine all aspects of the performance of filtration equipment—not just efficiency. Life cycle and pressure drop are evaluation factors that are far more important criteria than first cost. This is because some of the more expensive filter configurations, such as the minipleat, cost a great deal more initially than comparably efficient filters. Yet they are far better values because they last much longer than their competitors and operate at much lower pressure drop.

Use Gas Phase Filters Where Necessary

Gaseous/molecular filtration of chemical constituents is more difficult to select and apply to commercial buildings. Although industry consensus test methods for gas phase media and filter cartridges have been developed, they have only recently been published and have not been widely adopted by manufacturers. The filter cartridges required for acceptable extraction effectiveness and capacity performance are large and bulky, heavy, restrictive to airflow, and expensive relative to commercial particulate filtration. The sorption media types currently available—activated carbon, treated carbon, and potassium permanganate treated alumina pellets—also require rigorous and costly maintenance. For these reasons, it may not be desirable or practical in many cases to retrofit gaseous filtration in existing systems unless deemed necessary by risk assessment evaluation.

Use Outside Help on Specialty Filtration
and Air Cleaning Systems

It is feasible to apply HEPA particulate filtration and total retention gaseous filtration, if engineering modifications are performed to the air handling system to accommodate for the physical and mechanical requirements of the filters. Building managers should seek professional guidance in selecting the high-end filtration equipment, especially gas phase filters. This is because the commercially available sorbents and their containment module configurations vary dramatically in performance, cost and installation rigor. Further, their chemical sorption performance in controlling a specific contaminant is dependent upon airflow and bed dwell time, the chemistry and concentration of the contaminant/agent, and the chemistry of the sorbent. Thus, sorption media selection, media bed depth, module face velocity, module media content combine to influence the overall air cleaner performance selection. This makes the application and selection of gas phase filtration equipment a specialized skill not widely available in the engineering community.

More working information about filtration and air cleaning technology, history, test methods, and application can be found in the filtration chapter earlier in this book. Further, the reader is directed specifically to Figure 8-9 in that chapter that details and cross-references MERV levels with older designation and filter descriptions. Several other useful documents are now available to assist the building owner in understanding the risk and the response to exceptional environmental challenges. NIOSH has available on their web site two worthy documents—one dealing with general recommendations, which is basically a declassified military document, and the other specifically dealing with filtration for chem/bio control. ASHRAE also has guidance available in their Reports of the Presidential Ad Hoc Committee on Building Security and Response to Exceptional Environmental Challenges, of which this author was a member. (See Resources/References Appendix for contact information.)

Chapter 10

OPERATIONS AND MAINTENANCE: AN OUNCE OF PREVENTION

The energy efficient building that is not designed with IAQ resiliency in mind is less forgiving of poor maintenance. Problems that were once "blown away" in the old leaky buildings are now retained and remain to amplify and haunt us.

To save energy, leaks in the building were plugged to reduce infiltration and exfiltration. Infiltration represented uncontrolled outside air, which caused dirt, drafts, and discomfort. Efforts to "tighten" a building and reduce infiltration not only cut the costs of conditioning this unwanted air, but added to occupant comfort. The "tight" building, however, has lessened an important incidental means of diluting indoor pollution and was far less forgiving..

In another energy cost cutting effort, natural ventilation was reduced and mechanical systems were changed to recycle more of the already conditioned air. Unless the recirculated air was filtered and cleaned sufficiently, the pollutants just kept coming around and around like old battered suitcases on a baggage claim carousel. As each new "flight" arrived, the baggage piled up. Microorganisms, provided happy breeding grounds by poor "ground control," were trapped and recycled in the mechanical ventilation system. As they recycled in the tight building, the concentration of pollutants increased.

Poor maintenance, particularly in the mechanical system, is repeatedly cited by authorities and investigators as a major cause of indoor air pollution. It is critical enough that ASHRAE has added chapters on Operation and Maintenance in their current edition of Standard 62.1 for commercial buildings. As further evidence of the important of operations and maintenance, ASHRAE has also developed a Standard that specifically and competently addresses this issue titled ASHRAE/ANSI Standard 180-2008 Standard Practice for Inspection and Maintenance of Commercial Building HVAC Systems. This standard has been published since the previous Edition IV of this book but its guidance is consistent with the content of this chapter.

INHERITED "ENERGY CRISIS" PROBLEMS

Our energy problems of the 1970s proved to be a double-edged sword, cutting at maintenance both ways. The energy efficient building has not only made poor maintenance embarrassingly obvious, but climbing utility bills continue to be a major reason for cuts in maintenance budgets. This is particularly true in public institutions where budgets tend to be more rigid. In 1979, the American Association of School Administrators (AASA) asked their membership, the nation's school superintendents, where they were getting the money to pay higher utility bills. 68 percent responded that the primary source of the needed energy dollars was taken from operations and maintenance (O&M) budgets. And in that O&M "pocket," personnel costs were the hardest hit. Fewer positions, along with traditionally lower salaries, can cripple the maintenance efforts of dedicated plant engineers and facility managers. This trend has changed little in the school market since then, as schools continue to be a major problem category of buildings experiencing IAQ problems. Because one of the early signs of HVAC system degradation is reduced humidity control, mold has become a serious problem to school administrators.

Ironically, poor maintenance increases utility demands; prompting more cuts in the maintenance. This vicious downward cycle has left its mark on the quality of air we breathe.

In addition to cuts in manpower, a slow-down in equipment replacement as well as low *first-cost* decisions have compromised the facility manager's ability to maintain a productive environment. By 1983, the "Maintenance Gap" in our nation's public schools was $25 billion. By 1989, the Education Writer's Association put the differential in maintenance at $41 billion and by 1991, it had reach $100 billion. A 1988 study of college and university facility needs determined that the cost of capital renewal and deferred maintenance had reached $70 billion. Even the more sophisticated institutions have Deferred Maintenance Budgets that become "parking places and wish lists" of projects that should have been funded immediately, such as leaking roofs.

Responding to the need to cut energy costs, mechanical systems have become more complex. Modern air conditioning and ventilation systems developed seemingly unlimited possibilities of combining pre-fabricated components into complex systems. Frequently, operations and maintenance (O&M) training has simply not kept pace with staffing and appropriate training. The absence of trained personnel to operate and maintain the more sophisticated equipment as designed compounds the problem.

THE HIGH COST OF NEGLECTED MAINTENANCE

Many problems associated with indoor air quality are directly or indirectly related to operations and maintenance. Short-term economies in the O&M area, which are apt to reduce the quality of indoor air, prove to be very costly in the long run. Rather than incurring modest O&M costs, organizations are increasingly laying themselves open for higher IAQ costs. Costs that could, and should, have been prevented.

Once maintenance is deferred to this extent, the cost to "catch" up is leveraged to new heights. It is costly in productivity, in human relations, and in litigation threats or realities. Beyond these soft costs are the hard costs of replacing and upgrading mechanical equipment; renovating or reconstructing envelopes; restoring system control; and generally fixing the broken building, which far exceed the original savings of deferred maintenance. The "if it ain't broke, don't fix it" mentality comes back to haunt us with a multiplier that too often exceeds the cost of the building.

We learned the hard way that many sources and the means to control those sources are well known and many are traceable to maintenance practices. Since the costs to treat IAQ problems can far surpass the price of preventive maintenance, a new emphasis must be placed on the role of maintenance in assuring a safe, productive environment.

When O&M work, relegated to a lower priority, results in poor indoor air quality, there are apt to be social/political costs as well as remedial costs. Management may have a difficult time overcoming resentment when occupants realize they have been feeling miserable for days, weeks, even months because simple maintenance and potentially routine procedures were ignored. Economic losses associated with absenteeism and productivity can pale when compared to the long run impact of negative employee attitudes towards a management that didn't care enough to provide a healthy environment. To cap it all, the employees tend to expect a rather exotic solution to SBS problems and may not readily accept the idea that a simple maintenance action will now solve their problem.

Poor maintenance can, of course, be viewed by occupants as poor management... even negligent management; perhaps legally negligent management. The administrative and financial burdens associated with potential lawsuits further tip the balance in favor of effective operations and maintenance.

OPERATION & MAINTENANCE:
THE KEY IAQ INGREDIENT

Few people would care to take a ride on an airplane that they knew had not been properly maintained; yet, we regularly subject ourselves and others to unnecessary health concerns due to inadequate building maintenance. It is simply impossible to be assured of quality indoor air in a modern facility without a quality maintenance program.

Most investigators and authors now agree that problem building causes fall into three categories: (1) maintenance and operations, (2) system design, and (3) contaminant load changes. Of these three areas, the most frequent and apparent is poor maintenance.

Quality maintenance requires an effective facility management program. Reactive maintenance that simply responds to problems is sporadic and out of control. The facility management approach has four essential elements:

1. Top management support and commitment to the facility program, especially to operations and maintenance;

2. Planned O&M procedures; including a sound PM (Preventative Maintenance) program;

3. Qualified and trained personnel; and

4. Fiscal resources necessary to provide the manpower and materials required.

Operating the building and its equipment to meet environmental needs requires good maintenance and conscientious custodial practices carried out according to manufacturers' schedules and directions.

Building managers frequently experience turnover in operations and maintenance staff. The prevailing need to train new O&M staff means many facilities, especially public institutions, are usually playing "catch up" and are ripe for indoor air pollution problems. In an attempt to respond to indoor air problems caused by poor maintenance, the expedient, and usually costly, answer has been to increase outside air intake. This knee jerk reaction, relying totally on ventilation, is rarely the best solution. Dilution can mask the existence of potentially dangerous materials that may have long term hazardous effects. Settling for chronic low level exposure will leave us far short of what can be done to improve the quality of indoor air. Removing contaminants at the source, whenever feasible, is safer and more cost-effective than wasting energy using unnecessary ventilation.

The damage, disruption and human concerns related to each IAQ episode far outweigh the costs associated with an effective program. A relative small investment that provides training in identification, protocol and corrective maintenance procedures can save big headaches and costs later. As discussed earlier, the best management approach is a proactive approach, and "proactive" translated into O&M language is preventive maintenance.

PREVENTIVE MAINTENANCE FOR QUALITY INDOOR AIR

As every facility manager knows, a well-planned preventive maintenance (PM) program can prevent small deficiencies from blossoming into major, costly breakdowns, repairs and replacements, which always happen at the worst possible moment. A simple thing like the routine oiling of bearings in a fan can prevent an unscheduled loss of make-up air needed to assure quality indoor air.

Preventive maintenance has always been economically defendable, for it can:

• reduce unplanned service calls and the associated loss of man hours;

• reduce the number of equipment breakdowns;

• cut down the need for replacement materials and parts;

• reduce operating costs;

- create a more effective work environment for maintenance person-nel;

- lengthen equipment life;

- increase energy savings; and

- lower life cycle cost and increase payback periods.

Responding to IAQ concerns, we can now add:

- fewer IAQ complaints and the administrative time required to re-solve them;

- a more productive environment;

- a decrease in absenteeism; and

- lower health cost.

- avoidance or reduction of legal risk and liability

Clearly, the argument for preventive maintenance has become even stronger. The threat of indoor air problems has made preventive mainte-nance a critical function of effective management.

Setting IAQ implicated PM priorities as well as determining their timing and sequencing should consider known contaminant sources, building use schedules and equipment needs. For instance, the majority of maintenance-based IAQ problems are centered in the mechanical sys-tem; so this area deserves greater emphasis. Also, PM timing recognizes that pest control, painting and other actions involving the use of volatile organic chemicals should be done while few, if any, occupants are in the area.

Whether tracked manually or by computer, PM procedures should describe the actions to be taken and the frequency required. The dates and times (start and finish) of completed actions, the installation of any new parts, and the person who did the work should be a matter of re-cord.

PM planning can incorporate IAQ concerns in a number of ways. PM procedures may be described in relation to; (A) probable sources of specific contaminants and the preventive measures needed, (B) needs re-lated to specific equipment/systems, (C) services required by a particu-lar component common to several pieces of equipment, (D) the facility, or (E) in relation to energy efficiency measures.

Preventive Maintenance: Specific Contaminants

Table 10-1, taken from the Anne Arundel County Public Schools' (AACPS) manual, *Indoor Air Quality Management Program,* lists the sources and nature of contamination for a group of pollutants: bioaerosols. The AACPS approach was developed within the context of a comprehensive IAQ program. This school system is to be commended for its early lead in developing a forward thinking program. It was an excellent program then and it still stands as an excellent example of asset management through applied maintenance.

Using the table helps develop awareness of sources of a particular contaminant and implies, or directs, that some measures be taken in relation to those sources. It is a list that could also serve to remediate a problem when the known cause has been narrowed to bioaerosols. Just as important, this list taken within the PM context can help avoid the necessity to use the more agonizing and costly identification and mitigation processes. Table 10-1 can also serve as a model for incorporating other known sources/controls, such as those listed in Table 5-2. This table is provided as a generic prototype that can be adopted to the owners needs or to specific aspects of individual properties. The ASHRAE Standard 180 also provides some excellent working models of charts that are not reproduced here for reasons of copyright. The reader is urged to develop and employ simple checklists to guide maintenance practices regardless of their source.

Preventive Maintenance: Specific Equipment/Systems

Table 10-2 takes a different approach and lists traditional PM tasks that have IAQ implications for specific items of equipment within the air handling system. It does so without reference to specific contaminants. Table 10-2 illustrates the way the traditional PM approach can be used to satisfy IAQ needs with a subtle shift in emphasis. In this way, IAQ can be woven into an existing manual, or computer-based PM program, without a major change in operations.

Until the O&M staff has had IAQ training, this may prove to be a more expedient approach, as the PM staff can do the work without acquiring an understanding of the underlying IAQ needs. This approach, however, should never be regarded as a substitute for O&M training. Nothing takes the place of the staff understanding why they are taking certain actions, an awareness of when things "just aren't right," or that a particular action did not bring the expected results.

Table 10-1. Presumptive Sources of Bioaerosols in the Indoor Environment

Source	Nature of Contamination
Cooling coil section of air handling unit	Amplification of microorganisms may occur on wetted surfaces of cooling coils and in drain pans.
Humidifier containing reservoirs of stagnant water	Organic dusts and debris are scrubbed from the air stream; microorganisms may amplify in water reservoir and on wetted surfaces of device.
Steam humidifier	Condensed water from improperly trapped devices may serve as a niche for amplification of microorganisms.
Air washer; water spray system in air handling unit	Microorganisms grow on wetted mechanical surfaces, in water reservoirs, and on porous substrates associated with these devices.
Fan coil units; induction units	These devices may serve as reservoirs for microbial contaminants; check for accumulated dust debris.
Filters in air handling and fan coil units	Where maintenance is poor, filters may function as reservoirs for microbial contaminants; amplification can occur if humidity is excessive or if filter is wet.
Porous man-made insulation in ventilation system	Dirt and debris are trapped in porous areas; can become microbial reservoir; if insulation is wet microbial amplification can occur.
Outdoor air	Outdoor bioaerosols are the ultimate source of most indoor contamination.
Wet materials and furnishings	Ceiling tiles, carpet especially with natural fiber padding, wicker furniture, upholstery, etc., may function as microbial amplification sites.
Relative humidity greater than 70% in indoor environment	The equilibrium moisture content in organic dusts and substrates may rise to the point where it may support the amplification of microorganisms.

Source: Anne Arundel County Public Schools

Table 10-2. Preventive Maintenance Tasks: Air Handling System

FREQUENCY

ITEM AIR HANDLING SYSTEM	PROCEDURE	WKLY	MNTHLY	QRTRLY	SEMI ANNUAL	ANNUAL
AIR PATHWAY	CHECK FOR AIRFLOW OBSTRUCTIONS		•			
	CLEAN BIRD SCREEN					•
	CHECK FOR OPERATIVE LOUVERS					•
	CHECK AND CLEAN UP DEBRIS AND RUBBLE					•
	CHECK FOR ODOR/CONTAMINANT SOURCES					•
	CHECK FOR MOISTURE SOURCES		•			
	VERIFY VENTILATION AIRFLOW RATES					•
	CHECK FOR COOLING TOWER PLUME					•
	CHECK FOR AND CLEAN CRITTER OFFAL			•		
	CHECK CROSS CONTAMINATION SOURCES					•
HEAT/ COOL/ VENT	CHECK FILTERS; CLEAN IF NEEDED; CHECK CONDENSATE PAN DRAINS	•				
	CLEAN OR REPLACE FILTERS	•				
	CLEAN COILS & SPRAY WITH DISINFECTANT	•				
	CLEAN & FLUSH CONDENSATE PANS WITH DISINFECTANT AS NEEDED			•		
	CHECK & CLEAN AIR INTAKES			•		
	CHECK OVERALL DUCT SYSTEM FOR CLEANLINESS, LEAKS, MOISTURE, COLLAPSE			•		
	CLEAN HEATING/COOLING COILS		•			
	CHECK & CLEAN FAN/BLOWER BLADES OF DIRT & TRASH BUILDUP		•			
	CALIBRATE CONTROL SYSTEM				•	
RETURN AIR FAN	CHECK & CLEAN AIR INTAKES		•			
	CHECK OVERALL DUCT SYSTEM FOR CLEANLINESS, LEAKS, MOISTURE, COLLAPSE		•			
	CLEAN FAN & BLOWER BLADES	•				
CABINET HEATERS	CLEAN & CHECK FILTERS AS NEEDED	•				
	CLEAN OR REPLACE FILTERS		•			
	CLEAN COIL AND SPRAY WITH DISINFECTANT		•			

(Cont'd)

Table 10-2. (Cont'd)

FREQUENCY

ITEM	PROCEDURE	WKLY	MNTHLY	QRTRLY	SEMI ANNUAL	ANNUAL
AIR HANDLING SYSTEM						
FAN COIL UNITS	CLEAN & FLUSH CONDENSATE PANS WITH DISINFECTANT AS NEEDED			•		
	CHECK & CLEAN FILTERS AS NEEDED	•				
	CHECK & CLEAN COIL WITH SPRAY AND DISINFECTANT			•		
	CHECK AND CLEAR DRAIN LINES			•		
CEILING FANS	CHECK & CLEAN FAN BLADES		•			
UNIVENT UNITS	CHECK, CLEAN OR REPLACE FILTERS AS NEEDED		•			
	CHECK & CLEAN COIL AS NEEDED			•		
	CLEAN COILS & SPRAY WITH DISINFECTANT			•		
	CLEAN FAN BLADES & HOUSING				•	
	CHECK AND CLEAR DRAIN LINES			•		
EVAPORATIVE COOLER	CHECK FILTER. CLEAN AS NEEDED	•				
	CHECK GENERAL FILTER CONDITIONS			•		
	DRAIN UNIT, CLEAN & FLUSH SUMP & STRAINERS			•		
	SECURE UNIT FOR WINTER					•
HUMIDIFIER	CHECK FILTERS. CLEAN SUMP AND STRAINERS			•		
	LUBRICATE FANS & MOTOR BEARINGS				•	
INCREMENTAL UNIT	CLEAN OR REPLACE FILTERS		•			
	CLEAN COILS AND SPRAY WITH DISINFECTANT		•			
	CHECK CONDENSATE PAN, CLEAN PAN & DRAINS, FLUSH WITH DISINFECTANT AS NEEDED		•			
	CHECK & CLEAN COIL			•		
	CHECK AND CLEAR DRAIN LINES			•		
ROOF TOP UNITS						
HEAT/ COOL/ VENT	CLEAN OR REPLACE FILTERS	•				
	CHECK OPERATION OF ROLL FILTER		•			
	CLEAN COILS & SPRAY WITH DISINFECTANT		•			
	CLEAN & FLUSH CONDENSATE PANS WITH DISINFECTANT AS NEEDED		•			
	CLEAN ALL COILS	•				
	CHECK CONTROL OPERATION				•	
	CHECK AND CLEAR DRAIN LINES			•		

Adapted from Indoor Air Quality Management Program, AACPS

Table 10-2 is an illustration of how a PM program can be modified with reference to a specific equipment/system.

Preventive Maintenance: By Specific Components

Condensation pans are an excellent example of a piece of equipment that needs attention wherever found. They are perennial sources of indoor air problems if not properly maintained. Unattended condensate pans offer a dark, moist environment that is ideal for biological growth. They should be periodically cleaned and checked to be sure they are draining properly. Whenever a non-draining condensate pan is found, it should be repaired to operate properly and treated with an algicide according to product labeling.

Preventive Maintenance: The Facility

Since the HVAC system is frequently cited as a source of contaminants *and* as the mechanism dedicated to providing a comfortable environment, IAQ PM attention is rightfully focused on this area. There are, however, critical concerns connected with the building envelope and non-mechanical operations. Cracks in the foundation or openings around drain pipes, for example, allow radon to enter facilities. Bioaerosols grow in wet insulation or carpeting. Drains in laboratories are frequently overlooked as contaminant sources. Inactive drains can dry up and allow back-gassing of sewer gas. Biological organisms find the sediment in drain traps an ideal location to gather and grow. Any location where water or sediment can stand dormant in the plumbing should be kept clean, flushed, and checked periodically.

A review of contaminant sources/control will suggest a series of building envelope and non-mechanical inspection PM measures that should be taken on a regular basis, such as the following: adjacent contaminant sources; cross-contamination possibilities; moisture/water sources; high traffic areas; building envelope penetrations; and pressure relationships.

Preventive Maintenance: Energy Efficiency and IAQ

Blaming the energy efficient building is all to frequently used as a "cop out" to cover poor maintenance or to avoid a diligent search for the real problem.

Increasing energy efficiency as we improve IAQ has merit and is feasible. Rather than layering IAQ requirements over energy needs, they should be regarded as integral parts of the product building operators need to deliver: a comfortable, productive environment as cost effectively as possible. Rather than competing goals, IAQ and energy efficiency can and should share a commonality of purpose.

Not all efforts to improve indoor air quality need to be done at energy expense. Some measures are energy neutral, or positive, including:

- checking duct linings, especially in areas of high humidity, or moisture, correcting any problems;

- insure that insulation is dry since wet insulation loses its thermal properties;

- exercising care that steam for humidification has no contact with boiler additives;

- removing partitions that impede critical air flow, or inserting transfer grills or in-the-wall fans in the partitions;

- exercising great caution in the use and storage of chemicals in operations, maintenance, pest control, kitchens, etc. and using less toxic chemicals wherever possible;

- sealing cracks and openings around basement drains and openings, especially in radon affected areas;

- positioning air intake grills to avoid the reentry of fumes from the building's own exhausts or other avoidable outside contaminants, such as fumes from delivery alleys, cooling towers, and loading docks.

Many factors that have a negative impact on energy efficiency also have an adverse effect on IAQ. Correcting such situations can improve energy usage and enhance the quality of the air. Measures include:

- poor maintenance of pulleys, belts, bearings, heating and cooling coils and other mechanical systems can increase resistance causing a decrease in air supply. Good overall maintenance improves both energy efficiency and IAQ;

- water damaged insulation, ceiling tiles, rugs and internal walls support biological growth. Wet materials nullify insulating properties. Correcting water source and replacement increases energy efficiency while removing sources of biological contaminants;

- restricting uncontrolled infiltration improves comfort and reduces the air conditioning (heating and cooling) load on the equipment;

- leaks in terminal boxes and valves reduce temperature control, cause occupant discomfort and waste energy;

- humidity control can reduce the likelihood of mold while contributing to comfort and energy efficiency;

- inadequate or poor filtration allows coils to foul but also impairs air flow horsepower and heat exchange capability.

- inappropriate service cycles of filtration can allow excessive pressure loss to build-up and waste blower energy;

- malfunctioning, or inappropriately set, controls for outside air can bring in too much or too little air. Admitting too much outside air can actually burden the heating/cooling system beyond its capacity to respond, which detracts from occupant comfort. Running a mechanical cooling system when outside air can supply cooling needs is wasteful and works against occupant comfort and health;

- shutting down ventilation systems during unoccupied hours does not, in most instances, affect the quality of indoor air as long as they are turned back on in time to create a comfortable healthy environment prior to occupant arrival; IAQ can be impacted if the controls are not properly linked to toilet or kitchen exhaust systems which may continue to purge the building with humid saturated unconditioned air. This malfunction may, in fact, demand more energy to bring the building back under proper control during peak periods; and

- frequent causes of combustion contamination are defective central heating systems in which the exhaust is not vented properly, or there are cracks or leaks in equipment. Defective or poorly maintained systems are less efficient and, at the same time, pollutant sources.

Clearly, energy efficiency and indoor air quality are not only compatible components of the indoor environment, but many enhancement measures are mutually advantageous.

Special PM Activities
In almost every facility there are special PM needs with IAQ implications that warrant careful attention. Concerns related to a university's electrical transformers and capacitors, as discussed below, is an excellent example.

Federal law required that the university inventory its transformers and capacitors by October 1, 1990. The law also stipulated that the equipment be labeled and that location information be supplied to the local fire department.

In complying with these federal requirements, the university recognized that the polychlorinated biphenyls (PCBs) contained in this equipment, particularly if a fire were involved, posed a serious threat. The university, therefore, took a number of other steps as well. The director of physical plant included provisions in the PM program to dilute PCB concentrations in transformers to below 50 ppm and, wherever possible, to dispose of the liquid or metal casing. A scheduled replacement program was initiated and, until such time as the PCB transformers could be replaced, steps were taken to block ventilation and floor drains to prevent toxic combustion products from reaching occupied areas in the case of a fire.

The O&M emergency preparedness plan also included specific provisions for handling PCB fires or spills. A log book, also required by law, was set up to track inspections of PCB equipment and to record the approved disposal of the material.

PUTTING IT ALL TOGETHER

The approaches and specific concerns discussed above offer a number of ways to be sure IAQ needs become an integral part of a PM program. An elementary knowledge of contaminants and a little common sense in conjunction with accepted O&M procedures, will make any of these approaches workable. Contaminant sources and controls and treatments discussed in earlier chapters provide a sound basis for identifying existing PM tasks that need greater IAQ emphasis.

By comparing traditional PM procedures with IAQ prevention measures, it will become evident that incorporating IAQ into PM program adds very little to the work load. In fact, the comparison procedure only serves to underscore the importance of preventive maintenance in maintaining a healthy productive indoor environment. Further, as one building owner stated in the Building Operation Magazine "Good IAQ requires continual discipline. The key word in the phrase IAQ is "quality," and quality implies continual betterment. Quality means meeting the ever-increasing expectations of the occupants. As these expectations increase, the IAQ program should evolve and improve as well."

Whether IAQ concerns are addressed through a new or existing PM program, or some type of comprehensive maintenance effort, maintenance items that must be included can be summarized into a rather global items check list:

[√] inspect and change filters in all HVAC equipment regularly;

[√] aggressively manage water/moisture incursion and accumulation.

[√] clean coil and drain pans on a regularly scheduled basis;

[√] perform maintenance on mechanical system on regular schedule;

[√] maintain and calibrate controls on scheduled basis;

[√] regularly inspect and clean or replace, as necessary, microbially contaminated components;

[√] specify cleaning agents, procedures and schedule particularly for use in occupied spaces; and

[√] regularly inspect and repair all building envelope leaks.

POOR MAINTENANCE BY DESIGN

The annals of facility managers' lore are full of stories about air filters under a stage that *never* got changed, fan coil and induction units so difficult or time consuming to disassemble that they are never cleaned, equipment crowded against walls so that parts that require servicing can't be reached, etc., etc. The failure on the part of designers or builders to recognize maintenance needs takes on new importance when efforts to maintain indoor air quality are considered. The "failure by design" features most frequently cited by investigators are:

- Air-handling units were designed and manufactured without access doors to all heat exchanger sections, especially down stream of the cooling coil which is the "wet" end. Access panels are too often removable only with dozens of bolts and many man hours.

- Small air-handling units and heat pumps were located in inaccessible spaces above ceiling tiles. In many units, it proved to be physically impossible to gain entry into mechanical components to change filters, clean coils, drain pans, and other internal surfaces.

- Access and service doors blocked by other mechanical components, such as steam lines.

- Air handlers with poorly designed drain pans, improper air handler slope, poor drain location, and inadequate trap design.

- Mechanical rooms so remote and difficult to reach that boxes of filters must be carried over, under, and through ductwork.

- Filter sections located in remote ductwork without service doors.

- Spray humidifiers located immediately upstream of final filter banks or upstream of porous insulation or sound attenuators.

- Return air louvers in the mechanical room blocked by other equipment to disallow testing and balancing.

- High efficiency filter banks installed without any pressure gauges to indicate change cycle.

- Air-handling units located in rooms or plenums so confining or so close to walls, that human access was all but impossible.

- Filters installed in retainer or side load systems without proper functioning seals and gasketing.

The indoor air problems related to design extend beyond maintenance concerns and are discussed more fully under controls and treatment.

Poor maintenance by design could be largely avoided if the "resident experts" were only consulted. Building owners are missing an excellent opportunity for valuable input if they don't encourage architects and engineers to talk with their facility managers, directors of maintenance, plant engineers, operations supervisors, etc. These people have the hands-on information about what works and what doesn't in existing facilities. They are also in a position to foresee difficulties in contemplated designs.

Poor maintenance by design is supported, and compounded, by some rather common fallacies, ignorance, or "easy-way-out" rationalizations.

MAINTENANCE FALLACIES THAT WORK
AGAINST PRODUCTIVE ENVIRONMENTS

"Tight" Buildings Cause IAQ Problems

This is a great excuse to ignore leaks in the building envelope. The repeated use of the term "tight" building in referencing indoor air problems suggests it's a bad thing. It's not only that a leaky building allows uncontrolled amounts of air into a facility. It creates drafts and local thermal discomfort. A leaky building permits unfiltered air to enter a facility. Tightening a building envelope permits greater ventilation control and cleaner air. Provided the outdoor air brought in by the mechanical system compensates for infiltration losses, the tight building is one step closer to controlling the quality of the indoor air we breathe.

The Set and Walk Away Fallacy

Or, the "I've taken care of that and now I can forget it..." misconception. Buildings are dynamic. Changes in needs, programs, functions and use are almost constant. Occupants continuously disrupt the status quo. Ask the plant engineer about his or her day-to-day experience, the story is consistent: occupants can't leave anything alone. If they can reach it, they'll change it.

The Self-Sufficient BMS Fallacy

Building Management Systems (BMS) are sometimes viewed with trepidation by maintenance. Training is frequently inadequate, which leads to an assumption (perhaps a self-deluding one) that an BMS runs itself and doesn't need maintenance. BMS' major functions include the control of ventilation, heating and cooling. Poorly maintained BMS systems account for many indoor air problems and can actually squander energy.

"Everything Must Be Okay
Because They Haven't Complained," Fallacy

This comes under the, "If it ain't broke, don't fix it" rubric. Unfortunately, the fallacy equates "ain't broke" with a lack of complaints. IAQ problems, such as radon, or carbon monoxide that are invisible and non-odorous, do exist without immediate symptoms or complaints.

This fallacy is further supported by overburdened and undertrained staff. The problem is exemplified by the prevalence of staff who don't know how controls work or how to maintain them. The danger lies in the critical role controls can have in HVAC systems. As the HVAC "nervous system," they greatly influence the comfort and quality of indoor air.

Conversely, the problem is further exacerbated when the overburdened maintenance staff are called upon to "treat" complaints where no problem really exists.

New Construction/Design Fallacy

This fallacy has a number of aspects. First, there is an assumption by building operators that the design, because it is new, will satisfy health and safety concerns associated with the quality of indoor air. Unfortunately, many architects and engineers are still not well informed on IAQ needs with regard to design or materials specifications. Many SBS problems are associated with new construction.

Second, since the equipment was just installed and since it is covered by warranty, there is a fallacious assumption that it must be operating at design; therefore, the O&M staff can just coast. In an organization, the senior staff are often given the new building, which other operations staff view as a plum assignment as there is "hardly anything to do."

" It's-Running-at-Design" Fallacy

This may be the biggest culprit of all, because equipment seldom, if ever, operates at design. First of all, the O&M staff will reduce the operation of any piece of equipment to their level of understanding. Baling wire and chewing gum remedies are often preferable to admitting ignorance to the boss. But secondly, the systems may not have been properly commissioned to have even started out as intended by the design team.

These problems are compounded in still another way. Day time staff may assume around-the-clock controls are in place. Or, they may never have been informed as to the purpose of a certain switch or lever.

In some instances, operating personnel may be inclined to espouse adherence to standards and conditions they "ought" to follow; rather than own up to actual conditions. As a result, those who have done "midnight raids" on facilities frequently report that systems are not running as described.

Equipment age, inadequate or inappropriate maintenance and repairs all contribute to operations that stray from design specs. These problems are compounded by changing contaminant loads and equipment loads; so, even if equipment were operating at *design* once upon a time, chances are the current needs are no longer served. This affirms the need for facilities to be re-commissioned after major renovations, or timing milestones, such as every 5 or 10 years.

These fallacies only begin to reveal how many maintenance problems are rooted in misunderstandings, misconceptions, and the lack of the "right" information. All of which points to the vital role training holds in maintaining a staff qualified to do the job which is the last vital contribution that commissioning can bring to the successful design/construct process. Bulwer Lytton once observed "The pen is mightier than the sword." When it comes to achieving a productive cost-effective environment, this sage comment might be paraphrased, "The pen is mightier than the screwdriver." To put it another way, a well-trained O&M staff and a well thought out IAQ plan is widely regarded as the most effective "tool" in implementing an IAQ program.

Specific training suggestions are included in the later discussion of management procedures. No chapter on maintenance, however, would be complete without stressing the vital role training plays in implementing a solid PM program. Or, the critical role a preventive maintenance program has in assuring quality indoor air.

Chapter 11

MANAGEMENT PROCEDURES: THE SOFT SIDE OF IAQ SUCCESS

At exactly 5:13 am, the 18th of April 1906, a cow was standing somewhere between the main barn and milking shed on the old Shafter Ranch in California, minding her own business. Suddenly, the earth shook, the skies trembled, and when it was all over, there was nothing showing of the cow but a bit of her tail sticking up from the ground. IAQ is probably not the same force as the San Francisco earthquake, but it can have the same upheaval for managers who ignore its signals. Complaints are signals of imminent "earth tremors."

The way a complaint is handled can be the most crucial aspect of managing indoor air quality and warrants special attention prior to addressing broader management considerations. Whether the complaint is received by custodial staff, middle management, the CEO or someone talking to a board member at the company picnic, how that initial contact is treated can be critical.

COMPLAINT RESPONSE

Each IAQ episode exacts a price, human and economic. That price almost always far outweighs the costs of establishing an effective complaint response procedure.

A complaint that is lightly dismissed creates an emotional climate that is hard to overcome. A seemingly innocuous complaint that does not receive careful attention can grow into job dissatisfaction and lost productivity. Almost overnight, it can spread into a disgruntled work force, even a union dispute.

A complaint may actually be prompted by a cold or the flu. Or, a newspaper article, coupled with someone's active imagination and encouragement from the coffee lounge brigade, may start things going.

On the other hand, the symptoms may be from a very real building-associated illness. In either case, it can't be ignored.

When employees believe they are being forced to work in an area that is injurious to their health, they resent it. In the early days of IAQ pioneering, the Honeywell Diagnostics Group found psychological factors associated with SBS often involve employee distrust of management. That distrust can grow from a perceived no-action response by management to a broad range of employee complaints.

In an atmosphere of perceived disregard and distrust, emotions cloud the situation. By the time a serious investigation is undertaken, it is hard to sort the legitimate complaints from those made in the hysteria and anger of the moment. Recovering lost ground can be a long, painful, costly process.

Owner/tenant relationships can suffer the same consequence and the economic impact is apt to be more immediate and decisive. The tenant moves out. Word that the building is sick and the management is unresponsive spreads rapidly.

In too many instances, a "brushed off" complaint can lead to a very large public relations or legal problem even though the actual health problem may have been minimal. The air quality condition upon which the complaint is based may be minor, but the perceived indifference with which the complaint was received can explode into an issue beyond all recognition. Establishing procedures to respond to complaints is critical. A complaint response mechanism, because of its importance, frequently precedes any formalized IAQ program.

COMPLAINT RESPONSE PROCEDURES

Structure and Strategy
Step 1. Establish a complaint response procedure.
A critical aspect of establishing a complaint response procedure is briefing those receive complaints regarding who will receive the complaints, on attitudes and procedures. Every potential complaint recipient should know that the cardinal rule is to receive complaints with courtesy and careful attention. In other words, don't just hear, really *listen*. The person who complains may just want to be heard and to feel that someone cares. By listening and paying attention, a level of assurance is provided. Careful listening can also provide important clues that may

lead to the solution of an underlying problem.

Complainants view all discussions of indoor air quality and possible treatments through a veil of personal experience. That barrier cannot be pierced with talk of HVAC systems, or the like, until they have thoroughly exorcised their concerns, especially if the first complaint did not receive a personal response. They first must "unload" a litany of symptoms and any perceived neglect.

Formal notation of the complaint, and perhaps a formal interview later, not only gathers valuable data but sets the stage for more meaningful dialogue and interaction.

Step 2. Develop a complaint response tracking procedure.

A system for handling complaints of any kind should include a form for tracking and logging complaints. A complaint form and a log are essential management tools. In most instances, it is preferable _not_ to have the complainant fill out the form, as the question format may actually suggest answers. It is usually better to assemble the information through a brief interview. (More extensive interview procedures are discussed in conjunction with investigations in Chapter 4.)

If several occupants have reported the same condition, or if the source of the problem is relatively obvious, it is not necessary to go through a description of the problem on each form. However, remember that the objective is to be able to perceive patterns of similarity, including timing, location, or symptoms.

The IAQ Complaint Form

The form need not be complex, but there are certain elements that should be included beyond the obvious date, name and route of contact used by the person who reported the problem. A form, similar to the sample complaint form shown in Figure 11-1, should be available to record of the complaint and initial actions taken. The demographic information and the types of questions should include:

• Complainant, work area, date received.

• Nature of the problem. The gross symptoms, pattern and location and times of occurrence.

• Any noticeable odors or abnormal conditions with descriptions.

- The name of the person taking the complaint and the person accepting the complaint for follow-up, if different, should appear on the form.

- Space should be included for a *brief* description of possible follow-up/remedial action to be taken. When follow up detailed investigations or extensive remedial actions are needed, these reports should be attached to a copy of the complaint form for filing purposes.

Complaint forms are apt to be filled out by many different people in larger organizations. Any one with the authority to fill out the complaint, should be acquainted with all routing and disposition options.

It will defeat the purpose if complaints are allowed to sit on the corner of the desk for days. Getting the complaint to a central location promptly is a vital part of the process.

The Complaint Log

Once the complaint is received, it needs to be noted in a complaint log similar to the one shown in Figure 11-2. The log should provide space to note:

- the date received; who filed the complaint;
- who made the complaint and his/her work location;
- observed thermal environment and special conditions;
- the person designated for follow-up;
- disposition of the problem;
- if warranted, date and initials of any follow-up check or monitoring; and
- room for comments.

Step 3. Establish follow-up procedures.

Once the complaint is received and logged, it must be evaluated and some form of follow-up action taken. In many cases, the solution may be simple and quick. In others, a complaint may signal a problem that will require a great deal of detective work over an extended period of time.

If the investigation is protracted, it is important during this period that management make clear that complaints are taken seriously and that follow-up is real. This means that those who have complained *and their co-workers* need to be made aware that something is being done.

The Complaint Form

Building _____ Floor _____ Office/Room _____

Complainant _____ Date ___/___/___

Nature of work_____

In this work location since ___/___/___ Problems since ___/___/___

- -

Type of discomfort   Frequency of discomfort

Describe: _____   _____times day, wk., mo.

_____   Typical time _____

_____   Special conditions: _____

_____   _____

Environmental Conditions (circle, describe as appropriate)

too hot   too cold   too dry   too humid

noticeable odor _____   time _____

noise _____

lighting _____

ambience/furniture _____

other _____

Outside Conditions (circle, describe as appropriate)

sunny   partly cloudy   cloudy   rain   storm   snow

wind (strong/light)_____   outdoor temperature ___$^{\circ}$F

- -

Special or Abnormal Conditions, Comments

_____

- -

Complaint recorded by _____

Follow-up assigned to _____

Disposition (Date & initial all entries)

Is further follow-up [ ] or monitoring [ ] warranted?

Figure 11-1.

File # Assigned	Complaint Filed				Complaints/Conditions				Follow-Up			Comments
	Date	Bldg./Floor Office, Room	By	Complainant	Temp	RH	Vent	Other	Assigned To	Results	Initials	
001												
002												
003												
004												
005												

Figure 11-2. Complaint Log

In this case, there is nothing wrong with overt action, even when it creates a slight disruption the work area. In the case of an odor problem, it pays to do some "sniffing" while people are at work. Thus, they know investigators are on the trail of the culprit. Similarly, concern should be demonstrated. Ask questions and conduct the initial cursory examination of the ventilation system in the work place while it is occupied. This preliminary overt action is, of course, not just for show. Very often this relatively cursory investigation can provide early clues to the source of the problem.

Should a problem have medical consequences or require a detailed investigation, the diagnostic procedures will not be as simple, but the principle remains the same. Concerned workers need to know that the problem is getting careful attention and that measures are being taken to protect the occupants. If an investigation is underway, *it is not a secret!* It is best, therefore, to "tie a ribbon on it" and be open about what is being done. This avoids a flood of rumors and may be very useful in obtaining the positive cooperation of those who can provide valuable clues.

Step 4. Develop communication strategies.

Beyond the subtle and non-verbal communication potential available during preliminary investigations, a communications strategy must be an integral part of an emergency plan for any facility. It occupies a role no less important than meeting the challenge of a major power outage, the plan for evacuation, or terrorist activities. This proactive attitude gives credence to the expression "Chance favors the prepared mind." When conditions are less than ideal, effective communications can make the real difference between an inconvenience and a problem of enormous and never ending proportions.

Air quality problems may not require the speed of response of a boiler explosion or bomb threat, but communication with those involved is still a matter of high priority. Someone should have responsibility for the communications function within the emergency plan and should be fully aware of that assignment. This is not a task that can be left to "whoever is available." If the organization has the services of a professional public relations person, this assignment should be a part of the job description.

The collection and dissemination of accurate and timely information to the appropriate people is the cornerstone of a well-managed IAQ program—and the first place a troubled program "jumps the rails."

As an example of "jumping the rails" and to demonstrate how your audience (and the general public) perceives environmental issues, a colleague of the author relays this experience. His son, a junior in high school and a chemistry major developed a survey and an action petition. He contacted 100 residents in his neighborhood seeking their support for his appeal and asked for their signatures of support on a petition. He informed the citizens that their city government was widely marketing and distributing a potentially hazardous material. The hazardous material could cause respiratory failure if inhaled into the lungs, cause burns to the skin on contact, was such powerful solvent that it would dissolve many other materials, and could cause explosions when placed under pressure and heated. The student explained that the hazardous material in question was "di-hydrogen monoxide" and he sought their support in an effort to restrict the distribution of the hazardous material within their city. 99 of the one hundred citizens avowed their support against this "terrible material" and commented him for his civic concern. The one hold-out recognized the "hazardous" chemical to be H_2O—water. This illustrates several things about communications. First, people will accept half-truths and will be biased by the "twist" of how things are described and labeled (or mis-labeled). Secondly, people are simply not that knowledgeable about technical issues and they are ready to assume the worst about "threats to the environment." The message to building owners is that the 99 who signed the petition are typical of your audience when the "problem" strikes your building. Your communications policy, program, and training must take this into account.

The #1 Audience.

If there's a problem, those who occupy the facility must be the *first* audience. They need to be provided the facts, told what is going on and what is being done to rectify conditions. An audience that is well, *and accurately*, informed, greatly enhances the likelihood of keeping an indoor air problem confined to the facility. Failure brings about a much wider interest in the problem, including worker's families, the press and other outside groups.

Members of the #1 audience believe they have a real problem and, at least in their eyes, they deserve careful attention. They definitely should not feel their concerns are being ignored. Most facilities have at least one occupant who is never quite satisfied; so this is not always easy.

Communiqués should convey that someone is following up on the complaint(s), that management knows and cares about their concerns and will keep them informed.

Getting The Word Out.

When it has been determined that there is truly an indoor air problem, all the occupants of the facility should be made aware of the fact that management knows about it and is taking steps to solve it. How this is done will depend in great measure on the size of the organization, the occupants, and the seriousness of the problem at hand. In a school situation, for example, a potentially serious problem requires some communication with parents. In an office situation, such a broad effort may not be necessary. A simple posting or a "flyer" addressed to the occupants or tenant groups may suffice.

Communication to associated groups should be straight forward, stating facts as they are known, and seeing forth the steps that have been taken and are contemplated. This should include the investigatory procedures and remedial actions to the extent they are known. It is important not to wait until all actions are "cast in bronze" before bringing occupants up-to-speed.

Rumor Control.

Any organization with more than two people is fertile territory for rumors. And they will circulate at very nearly the speed of light. The most outlandish ideas will move as "actual facts" through any office, business, school or other facility where people get together. Rumors flourish in a "fact vacuum" making effective internal communications all the more vital. Rumors have a way of making the rounds even when the facts are easily available.

In the face of a serious indoor air quality problem, it may be desirable to designate an individual who can serve as a "rumor control center." Publicizing a telephone number where rumors can be checked and answers received will help curb rumors. It is absolutely necessary that the person delegated to handle rumors be kept fully informed of what is happening. If, through lack of current data, the wrong information is given out, the effort will be discredited and will backfire.

Progress Reports.

Planned progress reports can head off problems. Let the occupants of the facility know each step of the way what is being done to assess and "cure" the problem. If the solution will require a period of time, say so. Continued progress reports will help to control rumors and provide reassurance that management has not forgotten the affected individuals. This may be especially important in the case of public facilities.

Step 5. Develop Plans for Dealing with the Media.
In an area as volatile and emotionally laden as health problems caused by "unseen and uncontrolled villains" in our homes, offices and schools, the press plays a major role. Publicity has made more people aware that indoor air problems are a reality. Unfortunately, by repetition, over-coverage and failure to do sufficient homework, the press has been known to create a state of near panic where none was justified.

In some cases management has, in effect, allowed the press to diagnose a problem and prescribe a solution even before anyone could be certain that a real problem existed. This may not be the fault of the media, but a failure of management to communicate effectively with their own people or outside audiences.

Failure to acknowledge a problem when it exists and failure to explain what is being done to alleviate the problem leaves the way open for speculation and guessing from within and outside of the organization. Once false versions of what is occurring get started, they are virtually impossible to stop.

Projects do not always go smoothly or precisely follow a plan. Careful attention to communications throughout every phase provides the needed base of credibility if things start to go wrong. The key to successful passage through the communications mine field are openness, candor and adherence to the foremost rule of press relations: NEVER SPECULATE.

If problems arise, do not cut off the flow of information. Acknowledge problems while they are small and establish the fact that steps are being taken to deal with them. That way, hopefully, there will be no sudden revelation of a CRISIS.

When a serious (or sometimes not so serious) IAQ problem occurs in a public facility, it is fairly certain that the press will become involved. For those who have not dealt with reporters hot on the scent of a story, this can be an unsettling or frightening experience . But it need not be. Granted, it can be an intensely personal experience, especially the first time a person is misquoted by the press. So let's talk about the media and *you*.

The Media and YOU

There are a few basic rules to follow when dealing with the media that can take away much of the trauma and provide a much better chance that the newspaper or TV coverage of your problem will be accurate and fair.

Rule 1: Be accessible.

If you have a problem, don't try to hide it. It won't work! Select *one* person to be the spokesman for the organization and be sure that they have all the information they need. Your spokesman should make sure that press queries are answered promptly. Press people work on rigid schedules. If your message is going to help, it must be timely.

Rule 2: Be open and honest.

Tell them what you know and what you don't know. If an investigation is underway but no firm results are in, say so. Press people want instant answers to complex questions…they often know they can't get them, but they will ask anyway. Avoid giving preliminary information, because it may be wrong.

Rule 3: NEVER SPECULATE! NEVER SPECULATE!
(This is worth several rules.)

Speculation is telling something that you don't know. If the situation is less serious, speculation will amplify what could have been a minor problem. If the reverse is true, you have been guilty, or will be accused, of "covering up." Speculation is a sure way to lose when dealing with the media.

Answering a hypothetical question is a form of speculation. Don't succumb! Should a reporter's question begin with "IF," be on guard. Politely but firmly advise the reporter that you don't respond to hypothetical questions; then, turn the response to a point you want to make.

Rule 4: Admit you don't know.

If you don't know the answer to a question asked by a reporter, say so. If possible, find the answer and get back to that reporter with the information. If you have given an answer, and you find out you were wrong, contact the reporter with the correct information. You may be too late to get the right answer in the paper or on the air, but you will be helping your credibility with the reporter and it will pay off over time.

Rule 5: You're on the record.

Nothing you say to a reporter is *ever* "off the record." If you don't want to see it published, don't say it.

Rule 6: "No comment" is a comment.

If you say, "I can't comment on that because we don't have the facts yet," that is an answer. A flat "No comment" is almost a challenge and may leave the impression that you know, but are hiding the facts.

Rule 7: It's YOUR answer.

When being questioned in an interview situation, keep in mind that it is the reporter's question, but it is your answer. If you have a positive message that you want to get out, be sure to include it even if a leading question is not asked. It is perfectly all right to answer a question by saying, "I'll get to that in just a moment, but first I want to tell you about…"

Rule 8: Avoid pictorial amplification.

Being interviewed in front of a catastrophe, or with investigators decked out in masks or "moon suits," magnifies the story and frequently sends the wrong signal.

Rule 9: Take some comfort.

When all the news is bad and you seem to be all over the newspapers or the evening news, remember, most people will never hear about your problem. Sometimes it is a little hard on the ego, but it's reassuring to realize that everyone is self-centered enough not to pay too much attention to things that don't directly affect them. We are extremely aware of anything published about us, but everyone else is equally concerned with their own lives and may miss out completely on what grabs our attention. If you are taking what feels like a beating in the press, it generally isn't as bad as it seems.

MANAGEMENT BY COMPLAINT

The question that plagues management is how to determine whether a building is healthy or sick. Except for tobacco smoke, the threat cannot be seen and many pollutants are odorless. So what is the litmus test? Most of the literature starts with what Ylvisaker calls "the first and most obvious sign: tenant complaints."

While many owners and managers may get their introduction to indoor air problems through a complaint, relying on complaints leaves a lot to be desired. Using complaints as a barometer of indoor air quality poses two complications: (1) complaints do not necessarily mean the building is sick; and (2) the lack of complaints is no assurance the building is healthy.

In almost any facility at a given time, some occupants will voice complaints or evidence some of the symptoms typically associated with sick building syndrome. This natural occurrence can be blown into a full scale IAQ alert by an article someone found on office pollutants.

On the other hand, serious health consequences can emerge without evidence of symptoms or complaints long after the detrimental exposure has occurred. Problems associated with radon are a case in point.

If complaints serve as the primary trigger mechanism, an organization is automatically forced into a reactive mode.

Without a proactive approach to defuse unfounded complaints, the facility people are forever dashing about putting out "brush fires." This haphazard procedure places unexpected demands on manpower, usually when the organization can least afford to expend the time. Waiting "until the horse is stolen to lock the barn door" also leaves management in a legally vulnerable position. Failure to take appropriate measures, particularly if simple maintenance procedures would have avoided serious health consequences, is generally indefensible.

Once complaints are voiced, the longer it takes to identify the cause and to seek resolution, the more likely the performance, productivity and morale of everyone—even management—will be unfavorably affected. This may even trigger the "Princess and the Pea" reaction. After considerable complaints and attention are focused on the problem, no matter how many mattresses are added (or how much money is thrown at the problem) the "pea" is still "felt" and complained about by the occupants.

After the office is a-buzz because of a "contaminant crisis" is a poor time to start learning what the legal requirements are. Finding time to learn, for example, what community notification requirements exist under the federal Hazard Communication Standard only adds to the chaos. In the long run, the "wait-and-react" procedures are more time consuming, more emotionally draining and *more costly*. The high cost of reacting to complaints in such a fashion gives new meaning to the old adage: An ounce of prevention is worth a pound of cure.

The advantages of a proactive program include providing occupants a comfortable productive environment; enhancing management/employee relations or management/tenant relations; improving building operations and maintenance; and providing an increasingly important selling point for marketing property.

Managing indoor air quality is not unlike other management functions. It entails setting a policy, establishing someone to assume responsibilities, and developing/implementing a plan. The greatest time will always be consumed in initiating the program. A well-orchestrated plan, once in place, can be essentially self-sustaining.

DEVELOPING AN IAQ PROGRAM

To avoid falling in the response-action trap, the first step is to develop a basic policy and make necessary protocol decisions. Unfortunately, the impetus for developing an IAQ program is frequently an IAQ "experience" that no one wants to go through again.

POLICY

The key function of a policy statement is to demonstrate a strong commitment by the governing board and/or top management. The policy usually incorporates a statement of concern and commitment as well as broad goals and enabling language. More specifically, it will include:

- a statement of concern stressing the importance of indoor air quality and the need to maintain a safe, healthy productive environment for the occupants;

- a statement of commitment by the board/management;

- the broad goals—the conditions to be achieved and maintained;

- preliminary implementation considerations, such as;

— authorizing the position of IAQ manager, or designating such responsibilities to an existing position;

— delegating authority to that position within specified parameters, and

— requesting that an IAQ management plan be submitted for board and/or management approval;

• reporting requirements, which incorporate evaluation data and further recommendations; and

• baseline building quality data acquired as a part of proactive monitoring.

If the policy action has been prompted by a specific concern, the statement may make reference to this matter to reassure building occupants that top management is aware of, and committed to, resolution of the problem. For example, the direction to implement a plan may note respiratory problems being experienced by occupants in a particular area of the building and ask for a plan to respond to this need.

Other issues requiring top level support, such as new construction and purchasing procedures, might be addressed. If preliminary discussions reveal that a substantive shift in procedures or budget allocations will be required, a statement bolstering this change may be warranted. An examination of local IAQ conditions, for instance, may change operations and maintenance manpower needs in numbers, capabilities and training. Effecting this change requires resource allocations out of somebody's pocketbook.

Those giving up manpower and/or resources are inclined to create obstacles unless it is clear that top management is fully committed to the change.

The field of indoor environmental quality is into its fourth decade, yet policy on any external level has been slow coming. The OSHA announcement of intention to regulate in 1994 has finally been withdrawn, but the power of such documents is made obvious considering that the mere announcement of intent stopped tobacco smoking in public workplaces virtually overnight. There is guidance from the EPA and sporadic legislation at state level primarily on mold. Without firm guidance, the forward thinking manager will tailor his internal policy and planning to reflect the kind of information and documentation that may be expected by some future version of regulation that reflects the current state of the art.

THE IAQ MANAGER

The term, IAQ manager, sounds formal and formidable. It is used only to underscore the truism: If everyone is responsible, no one is. Someone must be in charge whether it's a new position, or responsibility is delegated to existing personnel.

An effective IAQ manager needs to have technical expertise and leadership/communications skills. The available choices run the gamut from a truly fine engineer with limited communications skills, who understands an air handling system and can detect HVAC problems at a glance, to the eloquent leader, who wouldn't know a bioaerosol if it bit him. The ultimate answer obviously will vary with the organization's needs and other engineering or communications capabilities on staff.

Although an IAQ manager's job description needs to be situation specific, it typically will include responsibility for: (a) setting up and implementing an IAQ program; (b) establishing and maintaining IAQ records; (c) assessing IAQ needs, overseeing surveys and inspections; (d) making recommendations in conjunction with other facility personnel and/or a diagnostic team; (e) implementing approved recommendations; (f) writing specifications, procuring services and equipment; (g) overseeing installation and operation of the equipment; (h) planning and implementing internal and/or external communications strategies; (i) monitoring and evaluating the program's effectiveness; and (j) routinely reporting progress to board/management.

The IAQ manager's job description, qualifications and organizational placement will tend to flow from the tasks central to the job. Typically, those tasks will focus on resolving current IAQ problems or developing/implementing a management plan.

MANAGEMENT PLAN

Because of the diversity and complexity of managing an IAQ program, a plan is needed to outline the tasks to be performed. A notebook accompanying the plan, which contains protocols and more detailed procedures, can help treat specific needs in greater depth.

Depending on the size and nature of the form, institution, or building, the plan is apt to vary in its formality, organization and presentation. An IAQ plan should be a living plan that moves with the times. If it sits on the shelf gathering dust, it was not responsive to needs at the time, or it was not flexible enough to meet changing conditions.

Most plans will include the following components, but not necessarily in the following order.

1. Person responsible for IAQ management. How, when, and where he or she can be contacted.

2. Purpose of the plan and how it is to be used.

3. Record keeping needs and procedures, such as
    —   records required by law (which may have an overlap with "Right to Know" records),
    —   complaint log and file of complaints/resolutions,
    —   employees' record of exposures
    —   findings from employee interviews; and
    —   investigation reports.

4. Compliance requirements, with reference to
    —   pertinent federal laws and regulations,
    —   applicable state statutes and local ordinances, and standards and guidelines; e.g., ASHRAE 62-2001 with addenda and interpretations.

5. Education and training procedures, which will cover
    —   staff briefings,
    —   the need for qualified personnel,
    —   procedures to document worker qualifications through training and experience, and
    —   any requirements for training attendance, certification, etc. of the organization's personnel.

6. Control and treatment procedures, including
    —   general guidelines with reference to specific pollutants as discussed in the resource notebook,

— safety procedures for O&M staff,

— purchasing procedures,

— time of use concerns,

— special protocols; e.g., paints, pesticides,

— temperature, humidity and ventilation parameters,

— new construction/renovation design, commissioning requirements, and

— storage of contaminants, disposal procedures.

NOTE: Some control issues may be listed separately in a
plan, such as purchasing and new construction, because of their
immediacy (new construction) or their importance to specific
segments of the organization. If, for example, the procurement
division is historically resistant to change, it may be necessary
to pull out—*and spell out*—exactly what needs to be done.

7. Procedures to authorize and select outside consultants to

— assist in establishing an IAQ program,

— train personnel, and/or

— conduct proactive or in-depth investigations.

8. Emergency preparedness procedures—reference to a separate
contingency plan.

9. Notification procedures including

— any required notification to local, state and federal agencies,

— notification to employees, including

   • postings, such as designated smoking or nonsmoking
     areas,

   • handouts that notify occupants of planned painting,
     spraying

   • information regarding exposures that have occurred,
     and

— notification to the community as required under the Hazard
Communication Standard.

10. Complaint response procedures

— filing complaints,

— tracking and logging, and

— follow-up.

11. Investigation protocol, including the
   — evaluation of healthy buildings,
   — interview forms, and
   — longer forms for data gathering to support in-depth diagnos-
      tics.

12. Communication strategies to keep employees and the community appropriately informed as to the
   — response to complaints,
   — investigations in progress,
   — remedial actions being taken, and
   — procedures for working with the media.

Concerns related to staffing, budget allocations and organizational procedures, developed as an adjunct to the plan, may be submitted as part of the plan or as a companion document.

The plan should be designed to satisfy the board's and/or management's concerns and to make the IAQ manager's job easier. The IAQ manager should be sure to put into the plan sufficient specifics to lend authority to his or her actions, but leave enough flexibility to meet unanticipated needs.

IAQ RESOURCE NOTEBOOK (Document! Document! Document!)

The most functional approach is to accompany a plan with a resource notebook. A loose-leaf notebook will permit easy revisions by section and the inclusion of new laws & regulations, articles, equipment data, etc. The loose-leaf approach facilitates revisions and helps keep the information current. The notebook can contain protocols and procedures in greater detail, such as painting or renovation protocols, response policies, and maintenance guidelines.

Information on each major contaminant or group of contaminants of interest can be assembled under dividers. Sections might also be devoted to purchasing guidelines, such as those available from American Conference of Governmental Industrial Hygienists (ACGIH), and new construction design requirements; e.g., ASHRAE 62-2001 and addenda, as well as guidelines for investigation from ISIAQ (the International Society of Indoor Air Quality).

MANAGEMENT CONCERNS

As the management begins to cope with indoor air concerns, several issues usually emerge for consideration. These issues include economic matters, particularly in relation to energy costs; legal issues; and such specific policy decisions as smoking bans and painting protocols.

ECONOMIC FACTORS

All management decisions include some assessment of the cost/benefit ratio. What will a recommendation cost? What will it gain? Just how cost-effective is it?

Central to the benefit side of the equation is the owner's concerns regarding tenant turnover and unoccupied space; or management's worries about losses due to lower productivity and absenteeism. One estimate comes from the USA Today who approximates the average cost of illness in the United States to be $572, although they also add that illness accounts for only about 25% of the "sick days" taken. They cite stress and personal issues as significant additional contributors to absenteeism. The financial losses and administrative nightmares to an owner of a sick building can be calculated. When the press reports a cost of $42 million to rectify the design/construction faults that lead to an uninhabitable facility, the costs to prevent such drastic action seem small in comparison. Costing out absenteeism is a little more difficult, as it may be clouded by other reasons for employee absence.

Lost productivity directly attributable to building associated health problems is hardest to measure. In 1989, respondents to a Building Owners Management Association (BOMA) stated that productivity would increase up to 21 percent if the air quality was improved. The BOMA survey reveals an estimate of the owner's perceived IAQ-associated productivity losses. At just 2 percent, the "IAQ" cost would run to $4 a square foot in an office building if total occupant cost runs about $200/sf/year. That cost doubles almost any preventive or remedial action needed to control most contaminants.

The Energy Cost Trade-off

A cost generally associated with improved indoor air quality is the utility costs for increased ventilation. Should energy costs climb in the coming years as projected, these costs will mount dramatically and ventilation as a control measure will come under greater scrutiny.

Careful management need not sacrifice energy savings for air quality. Energy efficiency and indoor air quality do not have an adversarial relationship, but rather are part and parcel of the comfortable, productive environment facility managers seek to deliver. Owners and managers need to understand the limitations of ventilation as a solution, as well as the ways engineers can address indoor air quality, which will satisfy the standard without automatically turning up the fan.

As energy prices climb, the need to recognize the options available and their cost-effectiveness will also grow. In the face of such a scenario, the EPA concluded and still espouses that the best control for any contaminant is at its source—not through dilution by ventilation.

LEGAL MATTERS

Potential legal costs become a part of any economic consideration. Legal fees and judgments in today's litigious society can weigh very heavily on budgets and management time. According to Mealey's Litigation Report for example, "Toxic" mold has cost the insurance industry at least $5 billion over the last few years resulting in the outcome that owners no longer have insurance coverage for this risk. The early warning of attorney/engineer Gardner, "Indoor Air Quality will be the toxic tort of the 90s," has come true into the new millennium. A clue to this fulfillment is the title of a litigators' seminar **M.O.L.D** with M designated as *money*; the O designated as *opportunity*; and L designated as *litigation*. One of the best reasons for developing a proactive IAQ management program is the ability to demonstrate a "reasonable man" position in order to assume the protective mantle that the doctrine affords. This does not position the building owner/manager to avoid litigation entirely, but it does attain an avoidance of a "negligence" decision and may even lead to early favorable settlement. It also can do much to avoid frivolous suits that are based upon improper facts about the building operation and maintenance that can be defended against with an active and documented management plan. The basic objective is to avoid the "lawsuit" which according to the 1906 Cynics Word Book is "a machine which you go into as a pig and come out of as a sausage."

SPECIFIC POLICY ISSUES

Some "policy" issues will be part of the policy statement, others may be part of an approved plan, and still others will be left to administrative judgment. How these issues are handled will depend on organizational size, management style and how much administrative "weight" is needed behind them. Concerns related to smoking and painting are excellent examples.

As demonstrated by the more recent trend for "No Smoking" buildings, the need for enforceable management decisions is central to environmental tobacco smoke (ETS) control strategy. Most ETS remedies require some control of occupant behavior through bans and designated smoking/nonsmoking areas. This process was substantially aided by the proposed OSHA ruling announced in 1994 even though it was never enacted, and has been withdrawn. The proposed ruling was a wide sweeping IAQ Regulation that involved all manner of reports, action plans, IAQ acceptance criterion, and all but banned smoking in the workplace. It received mountains of negative comments that eventually stopped it. However, the enlightened building manager would examine the document for potential positive actions that could be taken. It also provides an early warning about regulatory "thinking" should the document experience rebirth. More elaborate control policies and tactics are suggested in this book under the discussion of contaminant control methods. Such decisions, once made, must have the full force of the board/administration and top management to be effective. In radon affected areas, the medical, economic and legal implications take on added import as the radon/ETS synergism increases the likelihood of lung cancer significantly.

Some special protocols developed and circulated in advance, such as painting procedures, can very easily ward off a contaminant problem. The selection of the paint, the time of application, the advance notification to chemically sensitive occupants, and procedures to purge the facility prior to re-occupancy are all matters that can be resolved in advance.

SECURING CONSULTANT SERVICES

Much of IAQ management involves solid preventive maintenance and common sense. Technical needs can often be met with internal staff;

particularly, if the staff has received some IAQ training. There usually comes a time, however, when management needs outside assistance, especially in the planning stages or in diagnosing complex IAQ problems. When the time comes for outside help: be careful of the "snake oil salesman."

As in any specialized field, there are those who would like to profit from the client's ignorance. They can often be spotted by:

• their propensity to use IAQ jargon and meaningless statistics;

• a general reference to all the work they have done, but a failure to be specific;

• scare tactics; the "You-are-living-in-a-time-bomb!" approach; and "vacate the building" panic tactics; or emotional or emotional blackmail; e.g., letters of warning mailed directly to building occupants, or in the case of schools, to the parents or the press;

• a product, a new magical "black box," that will take care of *all* your indoor air problems; and/or

• a guarantee to discover the cause and correction of your indoor air problem(s) before ever stepping in the facility.

Chapter 4, Investigating Indoor Air Problems, discusses technical concerns related to retaining investigators and suggestions for using them effectively.

To avoid those who would jeopardize the organization's IAQ efforts, consultants, registered professionals or certified technicians should be hired in the same manner as any other professional: ask for and check credentials and references. Recognized trade and professional societies such as AEE (Association of Energy Engineers), AIHA (American Industrial Hygiene Association), and the IAQA (Indoor Air Quality Association) provide training and competency certification services for IAQ practitioners and professionals. The AEE "CIAQP" program (Certified IAQ Professional) provides guidance for selection of outside experts who are properly experienced and trained. Other training specialty groups provide training and competency certification for technicians in the HVAC and building service segments.

In the recent years at the end of the 90's and entering the first decade of the century, trade and professional organizations have proliferated, each offering specialty training and their own certification programs. Much of this feeding frenzy has been promulgated by rash of mold litigation and related legislation pending in many statehouses. Thus, the building manager faces an alphabet-soup of certifications and certifying organizations. This "certification war" has created a crowded arena of hungry entrepreneurs who are "experts" fresh from a two-day course and a certification exam. As with any procurement procedure, the management team must exercise extreme care and due diligence in the selection of professionals and contractors, depending heavily upon their demonstrated competence, documented experience, and a solid history of successful projects.

EDUCATION AND TRAINING

In addition to the communications strategies associated with complaint response discussed earlier in this chapter, the management has internal education and training responsibilities, including staff briefings and employee training. Some of these responsibilities may be fulfilled by the organization's own personnel; others may require an outside consultant.

Increasingly, many activities associated with an IAQ program are covered by laws and regulations. Ignorance of these provisions does not exonerate the individual or the organization from compliance. Legal precedents have been established for prosecution on the basis that the defendant " should have known. " Briefings and training, therefore, have become a key IAQ management function. Some training is specified by law, while additional training can afford individuals and the organization protection against costly mistakes or criminal liability. IAQ is a rapidly emerging field of concern at the federal and state levels; so checking with state agencies periodically is a good practice.

Staff briefings.

All staff need to know the purpose behind the IAQ program and the general procedures for implementation. They should be given guidance in reporting problems in a context that does not "manufacture" IAQ concerns or conversely, push them out of perspective.

Laws, regulations and standards that have implications for all employees need to be a part of the staff briefings and made available to them. Such documents should be part of the resource notebook.

Middle management and supervisors need to know what is expected of their workers and exactly what steps should be taken in case of an emergency. They should be advised of operational procedures, requirements under the law and their respective roles. As noted earlier, particular care needs to be taken so they understand information dissemination procedures to staff, public and the press.

If unions are involved, the union leadership should also be briefed, as appropriate, on the IAQ program and procedures.

Employee training.

Specific "hands-on" training should be provided to prepare workers to perform their tasks safely and effectively. This training should encompass: (1) the safe use and storage of hazardous materials; (2) how to perform their duties to avoid indoor air problems; (3) how to deal with any IAQ problems that may arise; and (4) procedures to assure air quality.

As stressed in the maintenance chapter, preventive maintenance (PM) is the centerpiece of an IAQ program. A PM training program with appropriate support materials should be developed and conducted to assure effective program implementation.

COMMUNICATIONS IN IAQ MANAGEMENT

Extensive communications relative to indoor air quality is not a major issue until "THE PROBLEM" emerges. Communications then tend to take on a level of importance that may well overshadow the real IAQ problem.

But, to limit concern for communication to times of crisis is a serious mistake that can have costly consequences. A continuous flow of information plays a vital role in any ongoing program to insure the quality of the indoor environment.

Effective IAQ management means that key people must understand and fully accept their role in all phases of the effort. Effective communications is the tool that helps to achieve that understanding and becomes as much a part of successful program as the wrench and screwdriver.

IAQ information needs to be presented to management in a straight forward manner. Burying people with detailed discussions of bioaerosols, VOCs or information about specific pieces of equipment will only confuse the issue. A general discussion should focus on what can be done, the problems that can be avoided, dollars that can be saved, and the comfort as well as health protection that can be achieved.

As an important side benefit, it has been demonstrated over and over that effective internal communications is one of the most effective external communications tools. The staff talks to friends and the word spreads.

Effective communications at all levels allows an IAQ problem to be handled in an atmosphere that permits an orderly approach, without panic. Good communications requires a small investment of time and resources, but the management dividends that accrue can be enormous.

In summary, effective IAQ management must be adapted to the organization's needs and capabilities. Whether an organization becomes IAQ sensitive because it has a problem or wants to avoid one, the inescapable components of an effective IAQ program are the statement of commitment, the designation of responsibility, identification of the technical and financial resources to do the job, a carefully designed communications strategy and a workable plan, which includes an effective complaint response procedure.

Chapter 12

WHAT "THEY" SAY
External Influences on IAQ

There are many excellent sources of information and guidance and the sources are growing and expanding at a rapid rate. These sources offer owners and operating personnel a means of keeping abreast of this rapidly changing field.

Resources are available to the practitioner from international studies, reports and conference proceedings. Several agencies of the federal government are involved in indoor air quality and states are increasingly developing supportive materials and some have related legislation pending, particularly in the area of mold.

A large number of associations and trade groups now claim some role and authority regarding the issue of Indoor Air Quality. Several associations have been very active in relation to IAQ concerns, particularly the American Society of Heating Refrigerating and Air-Conditioning Engineers (ASHRAE) and their trade counterpart the Air Conditioning and Refrigeration Institute (ARI), and the American Conference of Governmental Industrial Hygienists (ACGIH) and their trade counterpart the Association of Industrial Hygienist (AIHA). Two organizations solely dedicated to IAQ concerns include International Society of Indoor Air Quality and Climate (ISIAQ) and Indoor Air Quality Association (IAQA). ISIAQ's membership is predominantly international researchers and members of academia. IAQA's membership is predominantly IAQ practitioners specializing in consulting or contracting. (*As an aid to the reader, the full roster of players is denoted in the Appendix on Resources and References along with contact information, including web sites. The reader is urged to contact the listed organizations as many of them have excellent reference materials posted on their web sites covered many aspects of the IAQ issue.*)

ASHRAE 62, VENTILATION FOR
ACCEPTABLE INDOOR AIR QUALITY

What Does It Say and Why?

It has been decades since the first edition of this book was published, shortly after the publication of the 1989 version of Standard 62. Since that time, ASHRAE Standard 62 has become the "thousand pound gorilla" of the IAQ issue. From a relatively insignificant standards development exercise, it has grown in importance to one of the prestigious and highest profile standards in the ASHRAE stable of documents. As evidence of the emerging significance and growing power of the Standard, the revision process of 62 since the 89 version has been highly politicized, contentious, and adversarial. It has dominated the attention of the Society, its leadership, and its standards development process. Because of the importance and influence of 62 to both the IAQ practitioner as well as building managers, this chapter will deal with some of the history of the Standard—how it became the document it is today. It is not the intent of the discussion to "teach" the reader the details of content or how to use and apply the specifics of the Standard. Rather, the motive is to identify the significant impact and influence the Standard will impose on both building design and long-term performance. In early IAQ Fundamentals classes, Shirley Hansen would tell the story about the little boy who asked his mother to tell him about "turtles." His mother responded by tell her son "Why don't you ask your father—he's a marine biologist. Her son responded "But, Mom, I don't want to know *that* much about turtles." Similarly, the intent of this discussion is to "tell you about Standard 62" but to not tell you more than you never wanted to know.

ASHRAE has had a dominant influence in establishing comfort standards in the United States for many years. The definition of sick buildings uses the 20 percent indicator from ASHRAE Standard 55, which defined comfort in terms of conditions to satisfy 80 percent of the adult occupants.

ASHRAE's predecessor, The American Society of Heating and Ventilating Engineers, adopted its first ventilation standard in 1895 with 30 cubic feet per minute per person (cfm/p) as its minimum rate. This was based on earlier research in the mid 1800's by Tredgold, a Welch mining engineer working with canaries to establish survival rates of ventilation in mines. Remember that the ASVE standard of

'95 was issued in an era of an "un-washed" society. The primary intent of the standard was to address comfort from an occupancy odor basis, just as the subsequent ASHRAE standards focused on odor control. In 1973, ASHRAE issued Standard 62-73, Standard for Natural and Mechanical Ventilation, which recommended ventilation rates for 140 space usage applications. It used a dual ventilation rate table for "minimum" and "recommended" with a few spaces designated as low as 5 cfm/p if they were small with low population and likely to be non-smoking space. This dual table approach survived into the 62-1981 revision, which also more widely applied the 5 cfm/p minimum under designated conditions. Those conditional restrictions were that the ventilation air employed would meet the NAAQS Standard (National Ambient Air Quality Standard); there was no smoking in the space; and the return air was treated with appropriate filtration and air cleaning. These caveat restrictions were soon overlooked, however. The Standard was published in the aftermath of the energy crisis of 70's. Concurrently, ASHRAE issued the first version of Standard 90 that almost immediately became federal law. This document referenced the ventilation standard and mandated the use of "lowest" number. Thus, the decade of the 80s experienced a large number of buildings constructed to a "maximum" of 5 cfm/p outdoor air for ventilation for purposes of energy management due to government mandate. The requirements for filtering, smoking, and clean outdoor air were entirely overlooked.

Soon after 62-81 was published, research results raised some doubts regarding the 5 cfm rate. One study found body odors persisted at 5 cfm, but were no longer perceived by people entering a room at 15 cfm. This 15 cfm rate lowered steady-state CO_2 levels to 1000 parts per million (ppm). U.S. Army studies revealed a 45 percent increase in respiratory infections among recruits in energy efficient buildings, which had only 5 percent outdoor air—not 5 cfm; but 5 percent, or 1.8 cfm per occupant. The older barracks with 40 percent, which equated to 14.4 cfm per occupant, were found to have much lower infection rates.

According to Janssen, chairman of the committee to revise 62-81 (reissued as 62-1989), these two studies were sufficient to convince the committee that outdoor air flow should not be less than 15 cfm per person. Further, the decade experienced the advent of "sick buildings"—buildings that made occupants ill, especially selected government building and schools that were built during the 5 cfm-maximum era. This gave rise to the early term "Tight Building Syndrome" in response to the adverse health effects occurring in buildings that were constructed to the new "tight" energy standards. Thus, in response to both the advances in the science as well as the adverse field experience, Standard 62-1989 was published with 15 cfm/person as the minimum ventilation level under all circumstances. Also the dual table from '81 was eliminated along with distinction between smoking and non-smoking space.

STANDARD 62-89—THE ORIGINAL CONTENT

ASHRAE 62-89 contained both prescriptive (Ventilation Rate Procedure) and performance (Air Quality Procedure) criteria. These approaches covered alternative pathways to comply with the intent of the Standard. The Ventilation Rate Procedure presented a table of required cfm rates per person based upon a variety of space usage types. The total rate was then based upon occupancy population with the default occupant density assumption being 7 person per 1000 ft². The usage of the specified tabular rate implied compliance on an "ipso facto" basis. The Air Quality Procedure allowed the selection of ventilation rates on a performance basis with the potential of lowering outdoor rates and related conditioning costs through the usage of filtration and air cleaning. These dual approaches have persisted throughout the revision process.

VENTILATION RATE PROCEDURE

With 15 cfm/person being the then minimum rate, the ventilation rates in "Table 2" of 62-89 ranged from the minimum of 15 cfm to 20 cfm for offices and conference rooms to 60 cfm in smoking lounges. Classrooms were designed the minimum of 15 cfm/student. High density spaces such as theaters and auditoriums were also designated at the minimum of 15 cfm.

The standard did allow for some averaging to compensate for disparities, diversity, and extreme shifts in population density and varying room volumes. For example, an interpretation allowed 7.5 cfm/student, if sufficient reduction of density and periods of space usage could be documented. This averaging allowed for a reduction in outside air to adjust for the ventilation requirements of the most demanding space. However, precise calculations were required to determine the "critical space" which demonstrates the highest ventilation percentage rate based upon occupancy. This critical space determination may have required much higher percentages of ventilation air to serve the specific air handling zone containing the critical space than would be supplied with an overall average occupancy.

Probably the greatest weakness in relying on a quantity of outside air for contaminant control is the uncertainty as to how effectively the air distribution serves the occupants. Since the air outlet and return air inlets are generally located at ceiling level in office buildings, the circulation may be short circuited and increased outside air will not have the desired effect. The Standard's Table 2 assumed "well-mixed conditions " and this issue was addressed by requiring a ventilation efficiency factor of 100% with the proviso that systems having less effectiveness would require a proportionate increase of outdoor ventilation air.

62-89 also had a provision on design documentation, which requires the designer to state the design assumptions regarding ventilation rates and air distribution. The standard states that the documentation "should be made available for operation of the system within a reasonable time after installation." Since the owner is ultimately responsible for the building's well-being, it may be prudent in today's litigious society to discuss these assumptions with the designer *prior* to installation and to assure that the documentation is kept available in case of subsequent need during the life of the building.

AIR QUALITY PROCEDURE

Recognizing the limits of outdoor air as an effective and energy efficient control mechanism, 62-89 retained the indoor air quality procedure from 62-81. This procedure did not specify the flow rate, but offered objective and subjective performance criteria, such as contaminant concentrations held below acceptable limits and odor acceptability as determined by an odor panel. This procedure provided direct control of indoor air quality through enhanced particulate and/or gas phase chemical filtration, but also posed a problem, as the standard offers little guidance on assessing, calculating or controlling pollutants to acceptable levels. Furthermore, acceptable limits have not been well defined for many contaminants.

In some instances, guidance for indoor levels was taken as 1/10 of the Threshold Limit Values (TLVs) as published by the ACGIH (1985). This has proven inadequate, thus, the standard cautions that 1/10 TLV may not provide an environment satisfactory to individuals who are extremely sensitive to an irritant.

Guidelines for some contaminants originating indoors are shown in Table 12-1.

Table 12-1. Selected Air Contaminants Originating Indoors

Contaminants	ppm	Time
Carbon dioxide	700 ppm above ambient	continuous
Chlordane	0.0003	continuous
Ozone	0.05	continuous
Radon	4 pCil[a]	1 yr. avg.

[a]This EPA recommendation applies specifically to residences and schools. ASHRAE recommends its use until guidelines for other facilities are published.

Recognizing that all outside air is not "fresh" air, 62-89 provided for reduced outdoor air flow when the quality did not meet federal air quality standards. The National Ambient Air Quality Standard (NAAQS) for short-term concentration averaging was shown previously in Table 4-5. The standard applied the NAAQS contaminant levels to indoors for the same exposure times.

When outside air does not meet NAAQS levels, the filtration system becomes a critical factor for make-up air and recirculated air. Should it become necessary to reduce outdoor air, more recirculated air will be required and the quality of filtration, contaminant sensing and control will become central to the process of maintaining healthy indoor air. In effect, poor quality outdoor air forces designers to use the indoor air quality procedure to comply with the standard. The realities are that the Air Quality Procedure is seldom used in new construction unless the outdoor air is consistently undependable and a great deal is known about the space occupancy and usage. The procedure is most valuable on renovation and retrofit work where the following circumstances are likely to apply:

- When outdoor air is polluted, of low quality and/or in non-attainment of the EPA NAAQS Standard;

- When there is operating knowledge about the space usage;

- When contaminants of concern are identified and quantifiable both in outdoor air and return air;

- When there is need for upgrading the ventilation system to current higher volumes;

- When the condition of the outdoor air is designated "hot and humid";

- When the capacity of existing refrigeration systems and air handling system will not tolerate additional outdoor air;

- When the building occupant population density is high and/or widely diverse requiring large quantities of outdoor air during peak occupancy or at peak times; and

- When previous contaminant levels, such as mold remediation, require the usage of enhanced filtration systems for purposes of ongoing contaminant control.

These conditions create a favorable opportunity to take advantage of the ability of the Air Quality Procedure to provide cost effective IAQ enhancement through filtration and air cleaning. The savings in energy alone will more than offset the more rigorous and costly compliance documentation required by the Standard.

"ACCEPTABLE" AIR QUALITY

The dilemma that has plagued professionals in the IAQ field is the definition of "acceptable" in reference to the indoor environment. Further, does "acceptable" mean the condition of the environment is "safe"? And acceptable to whom? The original Standard 62-89 defined *acceptable air quality* as having "no known contaminants at harmful concentrations... and with which a substantial majority (80 percent or more) of the people exposed do not express dissatisfaction." This level of satisfaction and acceptability (80 percent) clearly means that conditions do not have to be unanimously or universally applicable to the entire occupancy population. In other words, to comply with the standard the space does not have to be conditioned to meet the unique needs of the hypersensitive—or the perennial complainer. Rohles, et al, have explored developing a rating scale for indoor environmental acceptability that incorporates thermal, acoustical, lighting, and air quality constituents. They developed a rating system that enables a numerical loading that concludes in an Environmental Acceptability Scale based upon 1 through 6 with the latter being "very acceptable." There work was performed through perception sampling of university students and staff. Although, this concept has not attained wide acceptance, it has value in establishing that IAQ acceptability is a complex and multifaceted characteristic.

One alternative suggested in the original 62-89 is the use of an odor panel. The problem with this approach is the rapidity with which people become inured to an odor even when the source continues to present a health problem. A judgment of acceptability must be rendered in 15 seconds to be valid. Research has actually shown that odor and irritation are mutually inhibitory; therefore, the existence of one can lessen the perception of the other. An odorant, therefore, could be removed and the irritation could still increase.

One way to use the indoor air quality procedure would appear to be CO_2 monitoring. Since the 15 cfm rate is predicated on keeping CO_2 under its actionable level of 700 ppm above ambient, procedures to keep CO_2 below 700 ppm rise (and ventilation at an acceptable level) can be monitored. The indoor air quality procedure monitored for CO_2 levels can assure acceptable ventilation in a more energy efficient manner, using ventilation as needed; not as some fixed volume approach. By the same token, if CO_2 can be controlled in offices, the rate/person guidelines can be satisfied without dollars to condition unnecessary outdoor air. Occupancy odors are controlled by the gaseous air cleaner system along with the VOCs and other gaseous constituents. Higher efficiency filters can then control the particulate portion of the air stream. Field investigation and evaluation can target the constituents of concern and similar evaluations can document control and conformance to the intent of the IAQ Procedure.

An underlying problem with cfm/occupant and CO_2 as an indirect measure is that the cfm guidelines rest on *per occupant* conditions. The CO_2 and odor surrogates are rooted in historical concerns regarding pollution by occupants. For hundreds of years, the focus has been on body odors, tobacco smoke and "stuffy" crowded rooms. We are still fighting the same old battle, but the rules of war have changed.

Many of our pollutant enemies on today's battlefield are not occupant-related. Furniture will off-gas at much the same rate if there are 2 or 20 people in a room. Bioaerosols can amplify in moldy carpets without help from people. Operating copiers, even carbonless copy paper, will pollute at the same rate with a few or a crowd. Our needs have moved beyond the body/tobacco stage requiring some procedure for sensing and controlling contaminants not related to occupancy in an energy efficient manner. These factors laid the groundwork for the inevitable revision of Standard 62—and especially the method for determining ventilation rates that accommodates for building related contribution to indoor concentrations.

REVISION ACTIVITY OF ASHRAE STANDARD 62

The text of the original Standard 62-89 was ready for approval in 1985 and formal publication was delayed by a series of appeals and delaying administrative rigors. In spite of this, the three-fold increase in the minimum ventilation rate—from 5 cfm to 15 cfm brought about considerable resistance from the construction community. Generally, the mounting number of building failures had triggered the term "sick-building-syndrome" that invoked a concern for health effects in addition to comfort control. Concern was also mounting by special interest groups, such as the Tobacco Institute, the Formaldehyde Institute, the Homebuilder's Association and the Building Owners and Managers Association. The Standard had been adopted totally or in part by all national code-writing bodies and its power and significance was building. These mounting pressures led to the action by ASHRAE to form a standing project committee (SPC) to address major revisions of the Standard 62-1989. The SPC was charged with updating the science and technology of the Standard; to address weaknesses and incongruities that had been exposed through application; and to revise the text to code language. This would, theoretically, ease the codification of the new version and assure that the content was not misinterpreted by code writing bodies through editing. Their revision work successfully developed a revision draft, but not without considerable controversy.

The draft titled "62-R Ventilation and Acceptable Indoor Air" was written in mandatory language and included a number of basic changes in concept. The ventilation rate determination, for example, separated the building related contaminant sources from the occupancy related sources departing dramatically from the traditional occupancy based body odor/tobacco smoke model. The revision required a minimum particulate filtration level, mechanical ventilation in residences, and also mandated commissioning, start-up, maintenance, and operation practices. Stringent humidity control, source control, demand ventilation, and other IAQ prevention tactics were also introduced by the draft. One of the most controversial aspects of the revision was its redefinition of acceptability to go beyond mere comfort to include irritation and health effects. This precluded the acceptability of any tobacco smoking in the conditioned space because of being declared a known carcinogen by a cognizant authority. By the time the draft was readied for public review in 1996, considerable opposition had developed from a number of affected parties, including a number of ASHRAE members.

The first public review resulted in almost 80,000 comments, many of which focused on the complexity and difficulty of applying the document, as well as its mandatory approach. Homebuilders objected to the more extensive and stringent ventilation requirements for residential construction. Equipment manufacturers were opposed to the requirements for higher efficiency filtration in unitary air moving devices. Building owners were concerned about the mandatory inclusion of maintenance and operation practices in a design standard. The strongest comments were voiced by regulatory and health based organizations regarding the apparent heath aspect of the new draft's scope of acceptability.

In light of the overwhelming negative comments, the revision draft was withdrawn and the ASHRAE board redirected the SPC to focus on the current Standard 62-1989 by placing it under "continuous maintenance." This is an ANSI approved process that enabled the revision of the then current standard on an item by item basis through individual addenda, which then go through the public review and consensus process. The process also allows addenda to be instigated by affected parties outside of the project committee. This action also made the standard project committee a Standing committee, which keeps the group in place permanently through rotating balanced membership to deal with ongoing revisions and interpretations. At the same time, ASHRAE created a separate committee SSPC 62.2 to develop a document to cover residential aspect of ventilation. This enabled a committee membership that was focused on residential construction. They successfully completed Standard 62.2, which was first published in 2003 and republished in 2007.

The revision process resulting from the Society Board actions has been successful in dealing with the revision needs of 62-89. The Standard was reissued in 1999; 2001; 2004; again in 2007, and most recently in 2010. In each revision, a number of new addenda were included that addressed a number of the issues brought forth earlier in the 62-R version that had to be withdrawn. Thus, it is safe to say that the "continuous maintenance" aspect of the current SSPC 62 has been successful in correcting many of the deficiencies of the early version. The series of addenda also has incrementally converted the original 62-89 guidance document into a code language (or mandatory) document. This has made the document and the technology that it represents more acceptable, applicable, and usable by the code writing bodies and the code authorities. The SSPC has also published a Users Manual that provides valuable insight regarding the application, interpretation, and rationale of the actual Standard.

On the other hand, the conversion of the document to code enforceable language also implicitly makes the Standard a "minimum" performance Standard rather than providing engineering guidance for above normal building performance. The addenda revision process also poses a severe challenge to building owners and IAQ practitioners because the minimum "standard of care" posed by Standard 62 is a moving target that changes with the issuance of every new addenda making it a "work in progress". To students of the 62 committee and the ASHRAE process that have followed the development of the Standard, it is clear that many if not most of the leading edge concepts inherent in the rejected 62R have now emerged over time through the incremental process of continuous maintenance. This process continues keeping the Standard in constant motion, which is good because is being constantly updated but is bad because it is a constantly "moving target". For practitioners performing diagnostics and forensics, care must be taken to ensure what applicable version of the standard was in place for that building or that renovation. For the building owner and the design team, ASHRAE has available a subscription service that for a nominal fee will automatically update your standard as new addenda are published. You will also be placed on an advice list of pending public reviews and significant standards actions.

WHERE TO NEXT?

There are some common themes running through both 62 Standards. They are both written in code language, which now prevails as common practice by ASHRAE. They are both coming under the "continuous maintenance" status, which means that new addenda will be created in a continuum to keep the content current and to respond to new technology and practice. They both address the need to deal with the building as a holistic entity. This recognizes the need to deal with performance of the building or residence (including air quality) as a unique balance of interdependent components and systems, including the air pathway, ventilation—both supply and exhaust—pressure barriers, controls, system cleanliness, and life/safety issues. The documents also affirm the role and critical nature of operations and maintenance as the sustaining activity to support the long-term performance and asset value of the building. It is important to note that the documents are minimum performance requirements, meaning minimum code compliant but not necessarily optimal for performance. And further, neither document assures the attainment of acceptable indoor air due to varying source strengths and individual building characteristics. As confirmation of the significance of these trends and in recognition of the need to go beyond minimum code compliance, ASHRAE has published Standard 180-2008 dealing with building maintenance and the Indoor Air Quality Design Guide-2009 dealing with advanced IAQ design. Both of these documents are important additions to the libraries of both practitioners and building owners. The latter document has received high praise for its in-depth holistic treatment of all of the factors influencing acceptable IAQ. At the same time, ASHRAE published Standard 189.1-2009 on sustainability and simultaneously updated their Green Building Guide. Pending publications include Guideline 10 on the Inter-actions of IAQ Factors and the long awaited Standard 145.2 which is the Test Method for gas phase air cleaner cartridges.

The prudent building owner should pay specific attention to these documents because they will be constantly changing due to the continuous maintenance process. From now on, they will be in code language, which will immediately create a new standard of care and will enable a rapid integration into law. Further, don't be surprised if these documents make their way quickly into the hands of occupants, tenants, or employees and their lawyers. To be thoroughly and consistently up-dated, contact ASHRAE @ www.ashrae.org to order their subscription service on Standard updates.

GLOSSARY OF ACRONYMS AND ABBREVIATIONS

ACH	air changes per hour
ACM	asbestos containing material
BASE	building assessment survey and evaluation
BOCA	Building Officials and Code Administrators
BRI	building related illness
BTU	British thermal unit
CAA	Clean Air Act
cfm	cubic feet per minute
CFR	Congressional Federal Registry
CFU	Colony Forming Units
CIAQP	Certified IAQ Professional
CIT	Certified IAQ Technician
CO	carbon monoxide
CO_2	carbon dioxide
DCV	Demand Control Ventilation
E & O	Errors and Ommissions
ETS	environmental tobacco smoke
GC/MS	gas chromatography/mass spectrometry
HCHO	formaldehyde
HEGA	High Efficiency Gas Adsorber
HEPA	high efficiency particulate arrestor or air filter
HVAC	heating, ventilating and air-conditioning (system)
IAQ	indoor air quality
IARC	International Agency for Research on Cancer
IEQ	indoor environmental quality
LCCA	Life Cycle Cost Analysis
MCS	multiple chemical sensitivity
MERV	Minimum Efficiency Reporting Value
mg/m^3	milligrams per cubic meter
MMMF	Man Made Mineral Fiber

MSDS	Material Safety Data Sheet
MVOC	microbial volatile organic compounds
NAAQS	National Ambient Air Quality Standards
NESHAP	National Emissions Standards for Hazardous Air Pollutants
NO_x	Oxides of Nitrogen
O & M	Operation and Maintenance
O_3	Ozone
OSHA	Occupational Safety & Health Administration
Pa	Pascal
PAH	polycyclic or polynuclear aromatic hydrocarbons
PCBs	polychlorinated biphenyls
pCi/1	picocuries per liter; a measure of radon concentration
PEL	permissible exposure limit
PM	particulate matter
ppb	parts per billion
ppm	parts per million
ppt	parts per trillion
RCRA	Resource Conservation and Recovery Act
RH	relative humidity
RSP	respirable suspended particles
SARS	severe acute respiratory syndrome
SBS	sick building syndrome
SPC	Standard Project Committee
SRD	source ranking database
SSPC	Standing Standard Project Committee
Stel	Short Term Exposure Limit
SVOC	semi-volatile organic compounds
TBS	tight building syndrome
TDS	Time Dependent Sensitivity
TILT	Toxicant-induced Loss of Tolerance
TLV	threshold limit value
TSP	total suspended particulate concentration
TVOC	total volatile organic compounds
$\mu g/m^3$	micrograms per cubic meter
μm	micrometer (micron)
UVGI	ultra-violet germicidal irradiation
VAV	variable air volume
VOC	volatile organic compound
WEEL	workplace environmental exposure limit
WL	working level

Appendix B

GLOSSARY OF TERMS

The Indoor Air Quality issue crosses a number of scientific and technical disciplines. The purpose of this glossary is to define the most commonly used terms in simple language for the understanding by those not skilled in IAQ related disciplines. To maintain consistency, the following terms in common with ASHRAE 62 definitions use the terminology which appears in the standard.

absorption: the process of one substance entering into the inner structure of another.

acceptable indoor air quality: air in which there are no known contaminants at harmful concentrations as determined by cognizant authorities and with which a substantial majority (80% or more) of the people exposed do not express dissatisfaction.

activated alumina: Pelletized aluminum oxide used as gaseous sorbtion filter when treated with potassium permanganate.

activated carbon: Granulized carbon treated with high temperature to create high surface area for sorbing gaseous contaminants.

adsorption: the adhesion of a thin film of liquid or gases to the surface of a solid substance.

aerosol: Fine droplets of liquid carried in air, i.e., fog or mist.

air, ambient: the air surrounding an object.

air, exhaust: air removed from a space and not reused therein.

air, makeup: outdoor air supplied to replace exhaust air and exfiltration.

air, outdoor: air taken from the external atmosphere and, therefore, not previously circulated through the system.

air, recirculated: air removed from the conditioned space and used for ventilation, heating, cooling, humidification, or dehumidification.

air, return: air removed from a space to be then recirculated or exhausted.

air, supply: that air delivered to the conditioned space and used for ventilation, heating, cooling, humidification, or dehumidification.

air, transfer: the movement of indoor air from one space to another.

air, ventilation: that portion of supply air that is outdoor air plus any recirculated air that has been treated for the purpose of maintaining acceptable indoor air quality.

air-cleaning system: a device or combination of devices applied to reduce the concentration of airborne contaminants, such as microorganisms, dusts, fumes, respirable particles, other particulate matter, gases, and/or vapors in air.

air-conditioning: the process of treating air to meet the requirements of a conditioned space by controlling its temperature, humidity, cleanliness, and distribution.

allergen: A substance that brings on an allergic reaction by the human organism, i.e., pollen or mold spores.

anaphylactic shock: An often severe and sometimes fatal systemic reaction in a susceptible individual upon exposure to a specific antigen (as wasp venom or penicillin) after previous sensitization that is characterized especially by respiratory symptoms, fainting, itching, and urticaria.

arthralgia: neuralgic pain in one or more joints.

bacterial: Referring to single cell plants that bring on disease, i.e., Legionella that causes Legionnaire's Disease.

biohazard: Substance that can cause harm to living organisms, i.e., exposure to radioactivity.

biosol: A contaminant carried in air that is either live or is the product of life chemistry, i.e., spores, or bacteria.

black lung/brown lung: Respiratory diseases resulting from long term exposure to airborne particulates specifically coal dust and cotton dust respectively.

carcinogen: Substance which provokes cancer in a live organism.

chemisorb: to take up and hold, usually irreversibly, by chemical forces.

communicable disease: Illness which can be passed from one individual to another by contact with a causal agent such as bacteria.

concentration: the quantity of one constituent dispersed in a defined amount of another.

conditioned space: that part of a building that is heated or cooled, or both, for the comfort of occupants.

contaminant: an unwanted airborne constituent that may reduce acceptability of the air.

dilution ventilation: dilution of contaminated air with uncontaminated air in a general area, room, or building for the purpose of health hazard or nuisance control. (ACGIH, 1984)

dose: the amount of exposure undergone at one time.

dust: an air suspension of particles (aerosol) of any solid material, usually with particle size less than 100 micrometers.

dyspnea: shortness of breath; difficult or labored respiration.

ecological/environmental disease: Category of disease, often respiratory, brought on because of exposure to a specific space or location.

ecological: Referring to habitat or the environment where an organism spends the most time.

endotoxin: A toxic or poisonous substance present in bacteria and separable from the cell upon its disintegration.

energy recovery ventilation system: a device or combination of devices applied to provide the outdoor air for ventilation in which energy is transferred between the intake and exhaust airstreams.

epidemiology: a branch of medicine that investigates the causes and control of epidemics; all the elements contributing to the occurrence or nonoccurrence of a disease in a population; ecology of a disease.

ergonomics: the study of people adjusting to their work environment; the science of adapting working conditions to the worker.

etiology: the science of causes or origins; the causes of a specific diseases.

exfiltration: air leakage outward through cracks and interstices and through ceilings, floors, and walls of a space or building.

fatigue: physical or mental exhaustion; weariness; tiredness.

filters: The following terms apply to filters:
   DOP Rated—filters having efficiencies determined by DOP (dioctlphthalate) smoke which is .3 micron in average size.
   Electronic Air Cleaner—high efficiency filter using electrical charge to attract small particles to a collection plate.
   Extended media—particulate filter having an enhanced surface area by some fashion of convolution.
   Final filter—higher efficiency extended media filter usually protected by prefilter.
   HEPA—High Efficiency Particulate Arrestor or so called "absolute" filter used in medical or clean room applications having ultrahigh—50%, 90%, 95%, 98%, 99.99%, 999.97% and greater efficiency against .3 micron sized particles (the latter is also referred to as "ULPA").

High efficiency—having a high degree of effectiveness against high loads and/or small sized particles.

Hospital—high efficiency filters used in medical facilities having MERV 13-16 efficiencies.

Odor or gas phase—sorption filters using pelletized media such as active carbon or treated alumina or zeolite to eliminate gaseous contaminants.

Mini-pleat—extended media filters using a pleated paper matrix held in a rigid frame similar to HEPA filters.

Pre-filter—lower efficiency filter usually relatively inexpensive but relatively short-lived.

TAW—"throw-a-way" or disposable low efficiency panel filter.

electret media—a non-woven fabric matte made from polymeric fibers having an inherent or manufacturing enhanced electrical charge.

formaldehyde: (HCHO) gaseous chemical contaminant causing irritation of skin, lips, nose, and throat. Suspected carcinogen found in construction materials such as particle board, laminants, fabrics, and carpeting.

fresh air: Misnomer for make-up air brought into building for dilution and positive pressure. Seldom is make-up air "fresh" and can be a source of indoor pollution.

fumes: airborne particles, usually less than 1 micrometer in size, formed by condensation of vapors, sublimation, distillation, calcination, or chemical reaction.

fungus: Variety of life form which colonizes in moist dark areas and propagates through small airborne spores, i.e., mold.

gas: a state of matter in which substances exist in the form of nonaggregated molecules, and which, within acceptable limits of accuracy, satisfies the ideal gas laws; usually a highly superheated vapor.

half-life: Period of time during which 1/2 of potency dissipates, such as radioactivity.

hazard: risk, peril, jeopardy to which an individual is subjected.

hypersensitivity: Lowered susceptibility to exposure to certain contaminants, i.e., hay fever.

incident: An occurrence of exposure, a spill, or high incidence or episode of disease.

infiltration: air leakage inward through cracks and interstices and through ceilings, floors, and walls of a space or building.

inhalable: particles small enough to be inhaled, but large enough so they are not quickly exhaled.

Legionnaires' Disease: Respiratory ecological disease caused by water-borne bacteria called Legionella. Name refers to the first identified outbreak in a hotel in Philadelphia during an American Legion Convention.

lethargy: a condition of abnormal drowsiness or torpor; a great lack of energy; apathy.

malaise: a vague feeling of discomfort or uneasiness.

micron: (Micrometer) Unit of linear measure of extremely small particles equal to one millionth of a meter, or one thousandth of a millimeter (.00003937 in). Bacteria are measured in microns.

microorganism: a microscopic organism, especially a bacterium, fungus, or a protozoan.

mildew: The visible filamentous product of mold/fungus colonies or growth, usually rust and downy or cottony masses.

mitigation: Refers to the total cleanup of an incident, from the people exposed, to site cleanup.

mutagen: Substances which cause mutation of offspring through chromosome alteration.

myalgia: pain in one or more muscles.

mycotoxin: Potentially harmful or poisonous metabolites of fungus especially molds.

natural ventilation: the movement of outdoor air into a space through intentionally provided openings, such as windows and doors, or through nonpowered ventilators or by infiltration.

occupied zone: the region within an occupied space between planes 3 and 72 in. (75 and 1 800mm) above the floor and more than 2 ft (600mm) from the walls or fixed air-conditioning equipment.

odor: a quality of gases, liquids, or particles that stimulates the olfactory organ.

oxidation: a reaction in which oxygen combines with another substance.

particle size: Measured in microns, refers to average size of airborne particulate contaminants.

particulate matter: a state of matter in which solid or liquid substances exist in the form of aggregated molecules or particles. Airborne particulate matter is typically in the size range of 0.01 to 100 micrometers.

pathogen: Live airborne substance having the potential of causing harm to an organism, i.e., disease producing.

picocurie: (pico = one trillionth) a unit of radioactive nuclide in which 3.7×1010 disintegrations occur per second

plug flow: a flow regime where the flow is predominately in one direction and contaminants are swept along with the flow.

polyuria: excessive secretion of urine.

Pontiac Fever: Earlier version of Legionnaires' Disease prior to its identification. Outbreak occurred in Pontiac, Michigan.

presumptive: giving reasonable cause for belief; presumed.

radioactive: Particle exposed to and carrying some form of radiation.

radon: Radioactive gas found in soil. A natural by-product of uranium decay.

relative humidity: (RH) Ratio of amount of water vapor in the air relative to the greatest amount possible at the same temperature.

respirable particles: respirable particles are those that penetrate into and are deposited in the nonciliated portion of the lung. Particles greater than 10 micrometers aerodynamic diameter are not respirable.

sorption: Filtration technique for controlling gas phase contaminants by molecular attraction and retention.
 Absorption—attraction of molecules into liquid phase.
 Adsorption—attraction of molecules onto surface.

smoke: the airborne solid and liquid particles and gases that evolve when a material undergoes pyrolysis or combustion. Note: chemical smoke is excluded from this definition.

threshold: In the context of odors, the point at which perception is barely discernible. In the context of exposure, the point at which tolerance is exceeded.

tolerance: Ability of an organism to withstand exposure levels.

tonnage: Archaic term for sizing refrigeration and air conditioning load, having to do with melting ice. One ton is 12,000 BTU.

total suspended particulate matter: the mass of particles suspended in a unit of volume of air when collected by a high-volume air sampler.

toxicology: Referring to the study of toxic exposures.

toxic: of, affected by, or caused by a toxin; to cause a poisonous reaction.

toxin: Highly poisonous contaminant yielded by life chemistry, i.e., fungus.

urticaria: An allergic disorder marked by raised edematous patches of skin or mucous membrane and usually intense itching. Caused by contact with a specific precipitating factor either externally or internally (hives).

vapor: a substance in gas form, particularly one near equilibrium with its condensed phase, which does not obey the ideal gas laws; in general, any gas below its critical temperature.

ventilation: the process of supplying and removing air by natural or mechanical means to and from any space. Such air may or may not be conditioned.

virus: Single cell animal having pathogenic effect on the human organism. Usually sub-micron but carried in air on a viable particle larger than 1 micron.

volatile: Property of a substance to evaporate readily into air.

Working Level (WL): A unit of radon exposure, a weighted average of concentration of radon daughters.

Appendix C

AIR-CONDITIONING, HEATING AND REFRIGERATION INSTITUTE (AHRI)

RESOURCES
Air & Waste Management Association (AWMA)
www.info@awma.org

Air Conditioning Contractors of America (ACCA)
www.acca.org

Air-Conditioning, Heating, and Refrigeration Institute (AHRI)
www.ari@ari.org

American Conference of Governmental
Industrial Hygienists (ACGIH)
www.acgih.gov

American Counsel for Accredited Certification
www.acac.org

American Industrial Hygiene Association (AIHA)
http://www.aiha.org,

American Institute of Architects (AIA)
infocentral@aia.org

American Lung Association (ALA)
www.lungusa.org

American National Standard Institute (ANSI)
www.ANSI.org

American Society of Heating, Refrigerating and
Air-Conditioning Engineers, Inc. (ASHRAE)
www.ASHRAE.org

American Society of Testing Materials (ASTM)
      www.astm.org

Associated Air Balance Council (AABC)
      www.aabchq.com

Association of Energy Engineers (AEE)
      www.aeecenter.org

Building Owners and Managers Association International (BOMA)
      www.boma.org

California Department of Health Services
      www.dhs.ca.gov

Centers for Disease Control and Prevention (CDC)
      www.cdc.gov

Consumer Product Safety Commission (CPSC)
      www.info@cpsc.gov

Department of Energy (DOE)
      www.energy.gov

Health and Human Services (HHS)
      www.os.dhhs.gov

Housing and Urban Development (HUD)
U.S. Department of Housing and Urban Development
      www.hud.gov

Indoor Air Quality Association (IAQA)
      www.iaqa.org

Indoor Air Quality Association (IAQA)
      www.iaqa.org

Institute of Environmental Sciences and Technology (IEST)
      www.iest.org

International Facility Management Association (IFMA)
    www.ifma.org

International Society of Indoor Air Quality and Climate (ISIAQ)
    Info@isiaq.org

National Air Duct Cleaners Association (NADCA)
    www.nadca.com

National Air Filter Association (NAFA)
    www.nadca.org

National Cancer Institute
    www.nci.nih.gov

National Environmental Balancing Bureau (NEBB)
    www.nebb.org

National Institute of Occupational Safety and
Health & Human Services (NIOSH)
    www.cdc.gov/niosh

National Institute of Building Sciences (NIBS)
    www.nibs.gov
    nibs@nibs.org

National Institute of Standards and Technology (NIST)
    www.nist.gov

North American Insulation Manufacturers Association (NAIMA)
    www.naima.org

Occupational Safety and Health Administration (OSHA)
    Department of Labor
    Washington, D.C. 20210

U.S. Public Health Service (USPHS)
    www.usphs.gov

U.S. Dept. of Homeland Security (USDHS)
www.dhs.gov

U.S. Environmental Protection Agency (USEPA)
www.epa.gov

Sheet Metal and Air Conditioning Contractors'
National Association (SMACNA)
www.info@smacna.org

World Health Organization (WHO)
www.who.int

REFERENCES (reviewed and/or used in this book or prior editions)
ACCA. "Good HVAC Practices for Residential and Commercial Build-
    ings. A Guide for Thermal, Moisture, and Contaminant Control to
    Enhance System Performance and Customer Satisfaction." Draft.
    April 27, 2003.
ACGIH. *ACGIH Threshold Limit Values and Biological Exposure Indices for
    1986-87*, American Conference of Governmental Industrial Hygien-
    ists, Cincinnati (1987).
AIHA. "Field Guide for the Determination of Biological Contaminants
    in Environmental Samples." 1996.
Adams, J.D.; O'Mara-Adams, K.J.; Hoffmann, D. "On the Mainstream-
    Sidestream Distribution of Smoke Components from Commercial
    Cigarettes." *39th Tobacco Chemists' Research Conference*. Montreal,
    PQ. 1985.
Ahmed, T.; Marchette, B.; Danta, I.; Birch, S.; Dougherty, R.L.; Schreck,
    R.; Sackner, M.A. "Effect of 0.1 ppm NO_2 on Bronchial Reactivity
    in Normals and Subject With Bronchial Asthma." *Am. Rev. Dis.* 125
    (1982), 152.
Air Quality Research International, "Performance of PF-1 and Dupont
    Passive Formaldehyde Monitors in AQRI Test Atmosphere Expo-
    sure Chamber." Prepared for the UFFI Center, 1984.
Air Technology Labs, 548 E. Mallard Circle, Fresno, CA 93710 (209) 435-
    3545.
Alevantis, L. "Designing for Smoking Rooms," *ASHRAE Journal*, July
    2003, Pgs. 26-32.

Amman, H.M., and M.A. Berry, N.E. Childs D.T. Mage. "Health Effects Associated with Indoor Pollutants," *Managing Indoor Air for Health and Energy Conservation*. Atlanta: ASHRAE, 1986.

Anne Arundel County Public Schools, *Indoor Air Quality Management Program*. Annapolis: AACPS, 1989.

ANSI/ASHRAE Standard 52.1-1992, (now withdrawn) Gravimetric and Dust-Spot Procedures for Testing Air-Cleaning Devices Used in General Ventilation for Removing Particulate Matter, American Society of Heating, Refrigerating, and Air-Conditioning Engineers, Inc., Atlanta, GA.

Asbestos Abatement Industry Directory, Contact: NICA/NAC, Department 5300, Washington, DC 20061-5030; (703) 683-6422, FAX (703) 549-4838.

"Asbestos in Schools—A Special Report," *School Business Affairs*. December (1988), 26-42.

ANSI/ASHRAE 55-2010, *Thermal Environmental Conditions for Human Occupancy*. ASHRAE, Inc, Atlanta, GA

ANSI/ASHRAE 62-1-2010, *Ventilation for Acceptable Indoor Air Quality*. ASHRAE, Inc, Atlanta, GA

ANSI/ASHRAE 90.1-2010, *Energy Efficient Design of New Buildings Except New Low-Rise Residential Buildings*. ASHRAE, Inc, Atlanta, GA

ANSI/ASHRAE Fundamentals, Chapter 12, "Air Contaminants," ASHRAE Inc. Atlanta GA. (2005)

ASHRAE Indoor Air Quality Guide (2009) "Best Practices for Design, ANSI/Construction, and Commissioning," ASHRAE, Inc. Atlanta, GA.

ANSI/ASHRAE Standard 189.1-2009 "Standard for the Design of High Performance Green Buildings," ASHRAE, Inc. Atlanta.

ANSI/ASHRAE Guideline 0-2005, "The Commissioning Process," ASHRAE, Inc., Atlanta, GA

ASHRAE Guideline 10-2010 *"Interactions Affecting the Achievement of Acceptable Indoor Environments,"* ASHRAE, Inc., Atlanta, GA.

ASHRAE Guideline 29-2009 "Guideline for the Risk Management of Public Health and Safety in Buildings." ASHRAE Inc., Atlanta, GA.

ANSI/ASHRAE Journal. "Psychosocial Links to SBS." January, 2003, Pg. 6. ASHRAE, Inc, Atlanta, GA

ANSI/ASHRAE RP-961- Part I: "Protocol for Measuring Contaminant Emissions from Renovation and Construction Activities." ASHRAE, Inc, Atlanta, GA

ANSI/ASHRAE Standard S12.60-2010 "American National Standard Acoustical Performance Criterion, Design Requirements, and Guidelines for Schools" American Acoustical Society, Melville, NY.

ANSI/ASHRAE Standard 52.2-2007, Method of Testing General Ventilation Air-cleaning Devices for Removal Efficiency by Particle Size, American Society of Heating, Refrigerating, and Air-Conditioning engineers, Inc., Atlanta, GA. ASHRAE, Inc, Atlanta, GA

ANSI/ASHRAE Standard 62.2-2010 "Ventilation and Acceptable Indoor Air Quality in Low-rise Residential Buildings, ASHRAE, Inc., Atlanta, GA.

ANSI/ASHRAE Standard 145.1-2009 "Laboratory Test Method for Assessing the Performance of Gas Phase Air Cleaning Systems: Loose Granular Media" ASHRAE, Inc. Atlanta, GA.

ANSI/ASHRAE Standard 180-2008 "Standard Practice for Inspection and Maintenance of Commercial Building HVAC," ASHRAE, Inc., Atlanta, GA

ASTM. "Standard Practice for Evaluating Residential Indoor Air Quality (IAQ) Concerns." Draft, November, 2001.

Bahnfleth, D.R. and F.A. Govan. "Effect of Building Airflow on Reentry and Indoor Air Quality. " *Practical Control of Indoor Air Problems* (1987), 185-193.

Balmat, J.L. "Generation of Constant Formaldehyde Levels for Inhalation Studies." *American Industrial Hygiene Association Journal*, Vol. 46(6) (1985), 690-692.

Balmat, J.L., and Meadows, G.W. "Monitoring of Formaldehyde in Air." *American Industrial Hygiene Association Journal*, Vol. 46(10) (1985),578-584.

Bayer, C.W., and Black, M.S. "Capillary chromatographic analysis of volatile organic compounds in the indoor environment." *J. of Chromatographic Science*, Vol 25 (1987), 60-64.

Bayer, C.W.; Downing, C. C., "Indoor Conditions in Schools with Insufficient Humidity Control." Paper. IAQ '92, Pgs. 115-117.

Benda, George, "Filtration is the Key to a Resilient Building" Air Media Magazine Spring 1997, National Air Filter Association.

Benda, George, "INvironment, the Handbook of Building Management and Indoor Air Quality," published by Powers Educational Services, 1991.

Benda, George, "Beyond IAQ and Energy: Resiliency Engineering." Paper.

Benda, George, "Raising the Bar for the Certification of IAQ Professionals." Paper.

Bender, J.R.; Mullin, L.S.; Graepel, G.J.; and Wilson, W.S. "Eye Irritation Response of Humans to Formaldehyde." *American Industrial Hygiene Association Journal*, Vol. 44(6) (1983) 463-465.

Berglund, L.G. and W.S. Cain. "Perceived Air Quality and the Thermal Environment." *The Human Equation: Health and Comfort*, (1989), 93-99.

Bernard, J.M. "Building-Associated Illnesses in an Office Environment.: *Managing Indoor Air for Health and Energy Conservation*, (1986), 44-52.

Black, M. "Formaldehyde in Construction and Building Materials." Georgia Institute of Technology, Indoor Air Quality Symposium, Atlanta, GA. 1985.

Black, M., et. al. "Correlation of Wood Product Formaldehyde Emission Rates as Determined Using a Large Scale Test Chamber, Small Scale Test Chamber, and Formaldehyde Emission Monitor," Georgia Institute of Technology, Atlanta, GA. 1985.

Black, M., "Cooling Towers—A Potential Environmental Source of Slow-Growing Mycobacterial Species." AIHA Journal, March/April, 2003, Pgs. 238-242.

Brennan, Terry M., "Responding to and Preventing IAQ Problems in Schools." HPAC Engineering, September. 2001, Pg. 34-43.

Brennan, Terry M. and James B. Cummings, "Unplanned Airflows & Moisture Problems." *ASHRAE Journal*, November, 2002, Pgs. 44-52.

Brosseau, Lisa M.; Vesley, Donald, et al. "Duct Cleaning: A Review of Associated Health Effects and Results of Company and Expert Surveys." Paper, ASHRAE Transactions, 2000 Vol. 106, Pt. 1.

Burge, H.A. "Environmental Allergy: Definitions, Causes, Control." *Engineering Solutions to Indoor Air Problems*, (1988), 3-7.

Burge, H.A., "Characterization of Bioaerosols in Buildings in the United States." Paper, IAQ '92, Pgs. 131-137.

Burge, H.A.; Chatigney, M.; Feeley, J.C.; Kreiss, K.; Morey, P.; Otten, J.; and Perterson, K. "Bioaerosols: Guidelines for Assessment and Sampling of Saprophytic Bioaerosolsin the Indoor Environment." *Appl Ind Hyg* 2(5), 1987.

Burge, Harriet, "Sample Analysis and Data Analysis," University of Michigan Conference; *Assessing Bioaerosol Hazards in the Workplace*. September 1987.

Burge, Harriet, A., "The Fungi: How They Grow and Their Effects on Human Health," HPAC Heating/Piping/Air Conditioning, July, 1997, pgs. 69-74.

Burroughs, H.E., "Air Filtration, A New Look at an Existing Asset," Global Engineering Conference 94, Carrier/Heating, Piping, Air Conditioning Magazine/Green Building Council, 1994.

Burroughs, H.E., "Breathing Easier: How to Choose a Filter That Fits," Good Cents Magazine, May/June, 1995.

Burroughs, H.E., "Filtration: A Cause and Solution for IAQ." Chapter 6, Improving Indoor Air Quality Through Design, Operation and Maintenance. Edited by Milton Meckler, published by The Fairmont Press, Inc., 1996.

Burroughs, H.E., "Filtration: An Investment in IAQ," HPAC Heating/Piping/Air Conditioning Magazine, August, 1997.

Burroughs, H.E., "IAQ and Green Buildings Compatibility: Fact or Fiction," Engineering Solutions to Indoor Air Quality Problems, AWMA/EPA, 1997

Burroughs, H. E., "IAQ: An Environmental Factor in the Indoor Habitat," HPAC Heating/Piping/Air Conditioning Magazine, February, 1997.

Burroughs, H.E., "Improved Filtration and the Residential Environment: An Intervention Research Project to Determine the Relevancy and Efficacy of Particulate Filtration in Improving Air Quality Conditions in Residences," Engineering Solutions to Indoor Air Quality Problems, AWMA/EPA, 1997.

Burroughs, H.E., "Indoor Air Quality," Chapter 17, Energy Management Handbook, edited by Wayne C. Turner, published by the Fairmont Press, Inc. 1996.

Burroughs, H.E., "Keeping Hvacr Systems Ready Against Attack." RSES Journal, February, 2002, Pgs. 23-27.

Burroughs, H.E., "Particulate Filtration, A New Look With New Data," Engineering Solutions to Indoor Air Quality Problems, AWMA/EPA, 1995.

Burroughs, H.E., "Sick Building Syndrome: Fact, Fiction or Facility," Chapter 1, Guide to Managing Indoor Air Quality in Health Care Organizations, edited by Wayne Hansen, published by The Joint Commission on Accreditation of Health Care Organizations, 1997.

Burroughs, H.E., "Specifying Filtration: How to Provide Building Owners with Optimal Design Value in Filtration and Air Cleaning."

HPAC Engineering. November, 2003. Pgs. 40-47.

Burroughs, H.E., "Taking Action Against the New 'Threat'." Part One. RSES Journal, January, 2002, Pgs. 32-36.

Burroughs, H.E., "The Role of Filtration with Building Security in the Post 9-11 World: Is Filtration the Silver Bullet?" Presentation at ASHRAE Annual Meeting, June, 2003.

Burroughs, H.E., "The Role of Risk Assessment as a Mitigation Protocol for IAQ Causes," AEE Proceedings, 1991.

Burroughs, H.E. Barney "The Usage of Filtration as Fulfillment of Acceptable Indoor and Optimal Energy Management," Proceedings of the Association of Energy Engineers, World Environmental Engineering Congress, 1992.

CDC, "Update: Pulmonary Hemorrhage/Hemosiderosis Among Infants-Cleveland, Ohio, 1993-1996." Vol. 46 #2.

Cain, W.S. "Sensory Attributes of Cigarette Smoking." *Banbury Report No. 3: A Safe Cigarette*? Cold Spring Harbor Lab, (1980), 239-249.

Cain, W.S., and Murphy C.L. "Interaction Between Chemoreceptive Modalities of Odour and Irritation." *Nature*, Vol. 284, (1980), 255257.

Cain, W.S.; Leaderer, B.P.; Isseroff, R.; Berglund, L.G.; Huey, R.J.; Lipsitt, E.D.; and Perlman, D. "Ventilation Requirements in Buildings. Control of Occupancy Odor and Tobacco Smoke Odor." *Atmosphere Environment*, Vol. 17, 6 (1988), 1183-1197.

Cain, W.S.; Tosun, T.; See, L.; and Leaderer, B.P. "Environment Tobacco Smoke: Sensory Reactions of Occupants." *Atmospheric Environment*, 21, 2 (1987) 347-357.

Cain, William S. and Brian P. Leaderer. "Ventilation Requirements in Occupied Spaces During Smoking and Nonsmoking Occupancy." *Environment International* 8 (1982), 505-514.

Cain, William S., Jonathon M. Samet, Michael Hodgson, "The Quest for Negligible Health Risk from Indoor Air," *ASHRAE Journal*, July, 1995, pgs. 38-43.

Carter, C.M.; Axten, C.W.; et al, "Indoor Airborne Fiber Levels of MMVF in Residential and Commercial Buildings." AIHA Journal, November/December, 1999, Pgs. 794-800.

Chang, P., Peters, L.K., and Ueno, Y. "Ventilation Requirements in Occupied Spaces During Smoking and Nonsmoking Occupancy." *Environmental International*, Vol. 8. (1985), 505-514.

Clausen, G.H.; Moller, S.B.; Fanger, P.O. "The Impact of Air Washing on Environmental Tobacco Smoke Odor." In B. Seifert *et al.* (eds.):

Proceedings of *Indoor Air '87*, Berlin, Vol. 2, (1987), 47-51.

Clausen, G.H.; Moller, S.B.; Fanger, P.O.; Leaderer, B.P.; and Dietz, R. "Background Odor Caused by Previous Tobacco Smoking." *Managing Indoor Air for Health and Energy Conservation*, Atlanta: ASHRAE, (1986), 119-125.

Coker, Bill, "Gas-Phase Filter Evaluations for Indoor Air Quality Applications." *ASHRAE Journal*.

Commins, B.T., *The Significance of Asbestos and Other Mineral Fibers in the Environment*. Commins Associates, Pippins, Altwood Close, Maidenhead, Berks S16 4PP, England, 1985.

Committee on Passive Smoking, National Research Council, and National Academy of Sciences. *Environmental Tobacco Smoke Measuring Exposures and Assessing Health Effects*. Washington, D.C. 1986.

Committee on Passive Smoking. National Research Council, National Academy of Sciences. *Environmental Tobacco Smoke Measuring Exposures and Assessing Health Effects*. (1986), 245.

Congressional Research Service "Report for Congress: Environmental Tobacco Smoke and Lung Cancer Risk, November 14, 1995," edited by Redhead, C. Stephen and Richard E. Rowberg.

Cummings, J.B. and J.J. Tooley, Jr., "Infiltration and Pressure Differences Induced by Forced Air Systems in Florida Residences," ASHRAE Transactions, 1989.

Cummings, James B.; Withers, Charles R.; et al. "Field Measurement of Uncontrolled Airflow and Depressurization in Restaurants." ASHRAE Transactions: Symposia. Pgs. 859-869.

Cummings, James B. "Moisture Control by Means of Pressure Control."

Cutter, "Filtration and Positive Pressure: Key Elements in Making Buildings Safer from Biological and Chemical Accidents or Terrorism." IEQ Strategies, Vol 14 #12.

D'Angelo, W.C., Spicer, R.C.; and Mease, M. "Sophisticated Asbestos Removal: Occupied Buildings and Operating HVAC." *Nat. Asbestos Coun. J.* 3 (1985), 9-14.

Department of Defense, "Unified Facilities Criteria (UFC) DoD Minimum Antiterrorism Standards for Buildings." Standard, July 31, 2002.

Dieckmann, John; Roth, Kurt W.; Broderick, James, "Dedicated Outdoor Air Systems." *ASHRAE Journal*, March, 2003, Pgs. 58-59.

Dols, W. Stuart; Persily, Andrew K.; Nabinger, Steven J., "Indoor Air Quality Commissioning of a New Office Building." NIST IR 5586,

January, 1995.

Dorgan, Chad B., "The Link Between IAQ and Maintenance." ASHRAE IAQ '98 Proceedings. Pgs. 32-47.

Dorgan, Chad B.; Linder, Robert J.; Dorgan, Charles E., "Application Guide Indoor Air Quality Standards of Performance." Book, ASHRAE RP 853, 1999.

Ellringer, P.J.; Boone, K.; Hendrickson, S., "Building Materials Used in Concstruction can Affect Indoor Funal Levels Greatly." AIHA Journal, November/December, 2000, Pgs. 895-900.

Emmerich, Steven J.; and Persily, Andrew K., "State-of-the-Art Review of CO2 Demand Controlled Ventilation Technology and Application." NIST IR 6729, July, 2001.

Elovitz, David M., "Minimum Outside Air Ventilation In VAV Systems," Engineered Systems, March, 1997, pgs. 44-50.

Environment: The Danish Town Hall Study." Environmental International, 13 (1987), 339-349.

Eto, J.H. and C. Meyer, "The HVAC Costs of Increased Fresh Air Ventilation Rates in Office Buildings," ASHRAE Transactions, pgs. 331-345.

First, M. "Environmental Tobacco Smoke Measurements: retrospect and prospect." *European Journal of Respiratory Disease*. Vol. 65 (Supplement), (1983), 9-16.

Fisk, William J., "MEGA $Billions can be Saved in the U.S. with Better Indoor Environments." Indoor Environment.

Fisk, William and Rosenfeld, Arthur, "The Indoor Environment…Productivity and Health and $$$," *Strategic Planning for Energy and the Environment*, Fairmont Press, Atlanta, GA, 2005

Fleming, W.S. "Indoor Air Quality, Infiltration, and Ventilation in Residential Buildings." *Managing Indoor Air for Health and Energy Conservation*, (1986),192-207.

Foarde, Karin K., "Determine the Efficacy of Antimicrobial Treatments of Fibrous Air Filters." ASHRAE 909-RP Final Report, July, 1998.

Fortune, Lamont, "Commissioning: Getting it Right." Engineered Systems, April, 2001, Pg. 26.

Franck, Carsten, "Eye Symptoms and Signs in Buildings with Indoor Climate Problems (Office Eye Syndrome)," ACTA Ophthalmologica, 1986, pgs. 306-311.

Fung, Frederick, "Health Effects of Indoor Fungal Bioaerosol Exposure." ASHRAE IAQ Applications, Fall, 2003.

Gardner, Thomas F. "IAQ: Legal Trouble for the 90s." *Air Conditioning, Heating and Refrigeration News*, (March 5 1990), 50-51.

Gemelli, F., and Cattani, R. "Carbon Monoxide Poisoning in Childhood." *British Medical Journal*. 291: (1985), 1197.

Gots, Ronald E.; Pirages, Suellen W., "Toxic Mold." ASHRAE IAQ Applications.

Green, George H. "The Effect of Indoor Relative Humidity on Absenteeism and Colds in Schools." *ASHRAE Journal*, (Jan 1975), 57-62.

Grot, R. et. al. "Validation of Models for Predicting Formaldehyde Concentrations in Residence Due to Pressed Wood Products," NBS, NBSIRT85-3255, Gaithersburg, MD. 1985.

Halvorsen, Tom, "Tracking IAQ Problems Using Ultrafine Particle Counting." ASHRAE IAQ Applications, November 10, 2000.

Hansen, Shirley J., "Indoor Air Quality: Is Increased Ventilation the Answer?," Energy Engineering, Vol 86, #6, 1989, pgs. 34-47.

Harriman, Lewis G.,III, "Better Dehumidification for Commercial Buildings." HPAC Engineering, August, 2001, Pgs. 23-27.

Harriman, Lewis G., III; Lstiburek, Joseph; Kittler, Reinhold; "Improving Humidity Control for Commercial Buildings." *ASHRAE Journal*, November, 2000, Pgs. 24-32.

Harriman, Lewis G. III; Witte, Michael J.; et al, "Evaluating Active Desiccant Systems for Ventilating Commercial Buildings." *ASHRAE Journal*, October., 1999, Pgs. 28-37.

Haurahan, L.P.; Anderson, H.A.; Dally, K.A.; Eckmann, A.D.; and Kanerek, M.S. "Formaldehyde Concentrations in Wisconsin Mobile Homes." *Journal of the Air Pollution Control Association*, Vol. 35,(1985), 1164-1167.

Hayner, Anne M., "The Lawyers are Coming," Engineered Systems, November/December 1993, pg. 4.

Hazucha, M.J.; Ginsberg, J.F.; McDonnell, W.F.; Haak, E.O. Jr.; Pimmel, R.L.; Salaam, S.A.; House, D.E.; Bromberg, P.A. "Effects of 0.1 ppm Nitrogen Dioxide on Airways of Normal and Asthmatic Subjects." *Journal Applied Physiology: Respir. Environ. Exercise Physiol.* 54 (1983), 730-739.

Health Risks of Radon, National Academy Press, 2101 Constitution Avenue, N.W., Washington, D.C. 20418

Hendry, Robert J.; Bayer, C. W., et al, "The Development and Verification of a Full Scale Gas Phase Filtration Apparatus and Methodology."

Hodgson, M.J. and K. Kreiss. "Building-Associated Diseases: An Update." *Managing Indoor Air for Health and Energy Conservation*, (1986),

Ingebrethsen, B.J., and Sears, S.B. "Particle Size Distribution Measurements of Sidestream Cigarette Smoke." *Proceedings of the Tobacco Chemists' Research Conference*. Montreal, Canada, October, 1985.

Ivanovich, Michael G., "HVAC Systems for TB Control," HPAC Heating, Piping, Air Conditioning Magazine, September, 1997, pg. 7.

Johns Hopkins Center, "SARS." Biodefense Quarterly Magazine.

Kleinman, M.T. Bailey, R.M.; Linn, W.S.; Anderson, K.R.; Whynot, J.D.; Shamoo, D.A.; Hackney, J.D. "Effects of 0.2 ppm Nitrogen Dioxide on Pulmonary Function and Response to Bronchoprovocation in Asthmatics." *Journal Toxicologv Environ. Health* 12 (1983), 815-826.

Kowalski, W.J.; Bahnfleth, W. P., "UVGI Design Basics for Air and Surface Disinfection." HPAC Engineering, January, 2000, Pgs. 100-109.

Krafthefer, B.C., "Air Conditioning and Heat Pump Operating Cost Savings by Maintaining Coil Cleanliness," ASHRAE Transactions, 1987.

Krafthefer, B.C., "Energy Use Implications of Methods to Maintain Heat Exchanger Coil Cleanliness," ASHRAE Transactions, 1986.

Kumar, Satish; Fisk, William J., "IEQ and the Impact on Building Occupants." *ASHRAE Journal*, April, 2002, Pgs. 50-52.

Kuraitis, K.; Richters, A.; Sherwin, R.P. "Spleen Changes in Animals Inhaling Ambient Levels of Nitrogen Dioxide." *Journal Toxicology Environ. Health* 7 (1981), 851-859.

Lane, C.A. and J.E. Woods, T.A. Bosman. "Indoor Air Quality Diagnostic Procedures for Sick and Healthy Buildings." *The Human Equation: Health and Comfort*, (1989), 237-240.

Latko, Mary Ann, "CDC's Guidelines for Tuberculosis Control in Health Care Facilities," INvironment Professional, April, 1995, pg 1, 4-7.

Leaderer, B.P. and W.S. Cain, "Air Quality in Buildings during Smoking and Non-Smoking Occupancy." ASHRAE Trans. V89 Part A & B, (1983),Paper No. DC-83-11.

Leaderer, B.P., Cain, W.S., Isseroff, R, and Berglund, L.G. "Ventilation Requirments in Buildings—II. Particulate Matter and Carbon Monoxide from Cigarette Smoking." *Atmospheric Environment*, Vol. 18, No. 1 1984),99-106.

Levin, H. "IAQ-Based HVAC Design Criteria." *Engineering Solutions to Indoor Air Problems*, (1988),61-68.

Levin, H. "National Expenditures for IAQ Problems, Prevention, or Mitigation. EPA (LBLL 58694) 6/2005.

Levin, Hal, "Estimating Building Material Contributions to Indoor Air Pollution," Indoor Air '96, July, 1996, pgs. 1-6.

Levy, Doug, "Linking Smoking, Lower Parkinson's Risk," USA Today, February 22, 1996.

Light, Ed; Johanning, E., et al, "How Should Health Departments Treat Stachybotrys in Buildings." NEHA Conference, 1999.

Liu, R.T.; Huza, M.A., "Filtration and Indoor Air Quality-A Practical Approach." *ASHRAE Journal*, February, 1995.

Lstiburek, Joseph, "Moisture, Building Enclosures, and Mold." HPAC Engineering, December, 2001, Pgs. 22-26.

Lubart, J. "The Common Cold and Humidity Imbalance." *NYS Journal of Medicine*, (1962), 816-819.

Mazurkiewicz, Greg, "Clean Those Filthy Coils." ACHR News, March 27, 2000, Pgs. 1, 16.

McCluskey, Gayla, "Mold, Mold, Mold." The Synergist, April, 2003, Pgs. 6-10.

McGrath, J.J.; Oyervides, J. *"Response of NO_2-Exposed Mice to Klebsiella Challenge."* In: *International Symposium on the Biomedical Effects of Ozone and Related Photochemical Oxidants.* Lee, S.D.; Mustafa, M.G.; Mehlman, M.A. (Eds.). Princeton, N.J.: *Princeton Scientific*; (1983), 475-485.

McGrath, J.J.; Smith, D.L. "Respiratory Responses to Nitrogen Dioxide Inhalation." *Journal of Environ. Sci. Health Part A* A19 (1984),417-431.

McKinney, Todd T., "High Efficiency Air Filtration: The Underutilized Tool for IAQ Control." Skylines, February, 1995, Pg. 16.

Mealey Publications, "$5 Billion and Counting...." Mealey's Litigation Report.

Meckler, M. and J.E. Janssen, "Use of Air Cleaners to Reduce Outside Air Requirements," *Engineering Solutions to Indoor Air Problems.* (1988),130-147.

Mendell, Mark J., "Non-Specific Symptoms in Office Workers: A Review and Summary of the Epidemiologic Literature," Indoor Air, 1993, pgs. 227-236.

Meyer, B., and Hermanns, K. "Reducing Indoor Air Formaldehyde

Concentrations." *Journal of the Air Pollution Control Association*, Vol. 35, (1985), 816-821.

Meyer, W. Craig, "Controlling Legionella in Cooling Towers." ASHRAE IAQ Applications, Spring, 2000, Pg. 9.

Milam, Joseph A., "Protecting Your Investment: A Continued Focus on IAQ Promotes Productivity in Changing Office Space," INviron-ment Professional, January, 1997, pgs. 3, 11-13.

Mikulina, Thomas W. "Increasing the Priority of Comfortability." *The NESA Newsletter*, Viewpoint (Sept 1989), 4.

Miller, C.S., "Chemical Exposures: Low Levels-High Stakes." Science Editor John Wiley and Sons, NYC. 1998

Miller, C.S., "Toxicant Induced Loss of Tolerance: Mechanisms of Ac-tion of Additive Stimuli Addiction" 96; 115-139.

Miller, C.S., "Chemical Sensitivity; Symptom, Syndrome, or Mechanism of Disease: Toxicology 111, 69-86, 1996.

Miller, J. David and J. Christopher Young, "The Use of Ergosterol to Measure Exposure to fungal Propagules in Indoor Air," AIHA Journal (58), January 1997, pgs. 39-43.

Millar, J. Donald, "Killer Mold: The Myth that Won't Die." PathCon Protocol, Vol. 5, #1, May, 2002.

Millar, J. Donald, George K. Morris, and Brian G. Shelton, "Legion-naires' Disease: Seeking Effective Prevention," *ASHRAE Journal*, January, 1997, pgs. 22-29.

Model EPA Curriculum for Training Management Planners, Environmen-tal Sciences, Ind. and Georgia Institute of Technology. July 1988. Available through ATLIS. 601 [Executive Boulevard, Rockville, MD 20852, (301) 468-1916.

Molhave, L. "Volatile Organic Compounds in Indoor Air Pollutants," *Indoor Air and Human Health*, R. Gammage and S. Kaye (eds), Lewish Publishers, Chelsea, MI. (1985),403-414.

Muller, Christopher O., "Packed-Bed vs. Carbon-Impregnated Fiber Gas-Phase Air Filters for the Maintenance of Acceptable IAQ." Paper.

Mumma, Stanley A., "Fresh Thinking: Dedicated Outdoor Air Sys-tems." Engineered Systems, May, 2001, Pgs. 54-60.

Muramatsu, T.; Weber, A.; Muramatsu, S.; and Akerman, F. "An Experi-mental Study on irritation and Annoyance Due to Passive Smok-ing." *Int. Arch. Occup. Environ. Health*, Vol. 51 (1983), 305-17.

NADCA, "NADCA General Specifications for the Cleaning of Com-

mercial Heating, Ventilating, and Air Conditioning Systems."
    NADCA Standard.

NAFA. Guide to Air Filtration. National Air Filtration Association,
    Washington, D.C.

NAFA, "NAFA User's Guide for Application of ASHRAE Standard 52.2
    Method of Test." User's Guide, NAFA, 2001.

Nardell, E.A., "Environmental Control of Drug Resistant Tuberculosis
    in Industrial and Developing Countries," Healthy Buildings/IAQ
    '97, pgs. 301-314.

National Council on Radiation Protection and Measurements. *Evalua-*
    tion of Occupational and Environmental Exposures to Radon and Ra-
    don Daughters in the United States. Bethesda, MD: National Council
    on Radiation Protection and Measurements: NCRP Report No. 78,
    1984.

National Research Council, Committee on Non-Occupational Health
    Risks of Asbestiform Fibers. *Asbestiform Fibers: Non-occupational*
    Health Risks, National Academy Press, Washington, DC, 1984.

NIBS, *Asbestos Model Guide Specifications.* NIBS, 1015 15th Street, NW,
    Washington, DC; (202) 347-5710.

Nicholas, Stephen W., "Underrating HVAC Air Filtration Systems."
    ACHR News, July 5, 1999, Pgs. 20-21.

NIOSH, "Guidance for Filtration & Air Cleaning Systems to Protect
    Building Environments from Airborne Chemical, Biological, or
    Radiological Attacks." NIOSH Draft, September, 2002.

NIOSH, "Guidance for Protecting Building Environments from Air-
    borne Chemical, Biological, or Radiological Attacks." NIOSH
    Report, March 15, 2002.

Noma, Elliot; Berglund, Birgitta; Berglund, Ulf; Johansson, Ingegerd;
    and Baird, John C., "Joint representation of physical locations
    and volatile organic compounds in indoor air from a healthy and
    a sick building." *Atmospheric Environment* Vol. 22, No. 3. (1988),
    451460.

Owen, M.K. and D.S. Ensor, L.S. Hovis and W.G. Tucker. "Effects of
    Office Building Heating and Ventilation System Parameters on
    Respirable Particles," *Managing Indoor Air for Health and Energy*
    Conservation. (1986), 510-516.

Perl, Peter and Li Fellers, "Breathing Uneasy," Washington Post Maga-
    zine, June 22, 1997, pgs. 13-28.

Persily, A.K. "Control Technology and IAQ Problems." *Engineering*

Solutions to Indoor Air Problems, (1988), *51-59.*

Persily, Andrew K., "Myths About Building Envelopes." *ASHRAE Journal*, March, 1999, Pgs. 39-47.

Pitkin County Health Dept. "Carbon Monoxide Exposures at a Skating Rink." MMWR 35(27) (1986), 435-441.

Radon Epidemiology: A Guide to the Literature, Susan L. Rose, Program Manager, ER-73, Office of Health and Environment Research, Department of Energy, GTN, Washington, D.C. 20545; (301) 353-4731.

Reed, Cynthia Howard, "IEQ Performance Metrics: Concepts & Context Background." NIST/ASTM, October 14, 2002.

Repace, J.L. "A Quantitative Estimate of Nonsmokers' Lung Cancer Risk from Passive Smoking," *Environment Int.*, Vol. 11 (1985), 3.

Richters, A.; Kuraitis, K. "Inhalation of NO2 and Blood-Borne Cancer Cell Spread to the Lungs. "*Arch. Environ. Health* 36 (1981), 3669.

Rohles, F.H., Jr., J.E. Woods, Jr., P.R. Morey, "Indoor Environment Acceptability: The Development of a Rating Scale," ASHRAE Transactions 1989, volume 95, part 1.

Russell, H.L.; Worth, J.A. and Leuchak, W.P. "Carbon Monoxide Intoxication Associated With Use of a Gasoline-Powered Resurfacing Machine at an Ice-Skating Rink." MMWR 33(4) (1984), 49-51.

Sawyer, R., "Asbestos Exposure in a Yale Building: Analysis and Resolution," *Envir. Res.*, 13 (1977), 1146-168.

Schell, Mike, "Real-Time Ventilation Control." HPAC Engineering, April, 2002, Pgs. 57-81.

Seppanen, O.A.; Fisk, William J.; Mendell, Mark J., "Ventilation Rates & Health." *ASHRAE Journal*, August, 2002, Pgs. 56-58.

Shaefer, Greg and William Hamilton, "Opposition Burns Over EPA's Final NAAQS for Ozone and Particulate Matter," EM Magazine, August, 1997, pgs. 21-28.

Shanley, Elizabeth and Steven Pike, M.D., "Don't Fall Into the Response Action Trap," *School Business Affairs*, (December 1988) 23.

Skov, Peder and Ole Valbjorn. "The Sick Building Syndrome in the Office

Sorensen, A.J. "The Importance of Monitoring Carbon Monoxide Levels in Indoor Ice Skating Rinks." *College Health* 34 (1986), 185-6.

Spengler, J.D., and Soczek, M.L. 1984. "Evidence for Health Effects of Side-Stream Tobacco Smoke." *ASHRAE Transactions*. Vol. 84, Part 15 (1984), 781-788.

Spengler, John D. and Haluk Ozkaynak, John F. McCarthy, Henry Lee, Harvard University, *Report Summary of Symposium on Health Aspects of Exposure to Asbestos in Buildings*. Conference Proceedings also available.

Spengler, John D. and Haluk Ozkaynak, John F. McCarthy, Henry Lee. *Summary of Symposium of Health Aspects of Exposure to Asbestos in Buildings*. Boston: Harvard School of Public Health, 1989.

Sterling, T.D. and Sterling, E.M. "Investigations on the Effect of Regulating Smoking on Levels of Indoor Pollution and on the Perception of Health and Comfort of Office Workers." *Eur J. Respir Dis*, Suppl 133, Vol, 65 (1984), 17-32.

Stetzenbach, Linda D., "Microbial Contamination & Indoor Air Quality." Air Media (NAFA), Fall, 2001, Pgs. 6-10.

Straden, E. "Radon in Dwellings and Lung Cancer—a Discussion." *Health Physics* 38 (1980), 301-306.

Straus, David C. and Jay Kirihara, "The Indoor Microbiological Garden," BOM Magazine, August, 1996, pgs. 78-84.

Streifel, Andrew J., "Airborne Infectious Disease: Best Practices for Ventilation Management." HPAC Engineering, Sept., 2003, Pgs. 97-104.

Surgeon General's Report. *The Health Consequences of Involuntary Smoking*, Washington, D.C.: U.S. Public Health Service. 1986.

The Radon Industry Directory, Radon Press, 500 N. Washington St., Alexandria, VA 22314.

Thornburg, Don, "Using Life Cycle Cost (LCC) Analysis to Select the Best Value Air Filters." Annual Technical Meeting, American Filtration Society, Fall, 2002.

Tulis, Jerry J. and Wayne R. Thomann, "Fungal Contamination and Growth in Ventilation Systems," Engineered Systems.

Umbrell, Christine, "Mold. Creating a Scientific Consensus on a Hot Topic." The Synergist, April, 2003, Pgs. 34-38.

University of Minnesota, "Mycological Aspects of Indoor Environmental Quality." Airborne Fungal Glossary.

US Department of Labor, "A Brief Guide to Mold in the Workplace." OSHA.gov, 10/29/03.

US Environmental Protection Agency, "The Inside Story: A Guide to Indoor Air Quality." EPA Document # 402-K-93-007, April, 1995.

US Environmental Protection Agency, "Should you have the Air Ducts in Your Home Cleaned?" EPA-402-K-97-002, October, 1997.

US Environmental Protection Agency, "Indoor Air Facts No. 4 (revised): Sick Building Syndrome (SBS). April, 1991.

US Environmental Protection Agency, "A Brief Guide to Mold, Moisture, and Your Home." EPA-402-K-02-003.

US Environmental Protection Agency, "Mold Resources." EPA-402-K-02-003.

US Environmental Protection Agency, "Guideline for Design for Schools."

US Environmental Protection Agency, "The EPA's Building Assessment Survey & Evaluation (BASE) Study Data and Website Scheduled for Release." Paper.

US Environmental Protection Agency, "An Office Building Occupant's Guide to Indoor Air Quality." EPA Internet.

US Environmental Protection Agency, "Design Tools for Schools." EPA Internet

Vergetis-Lundin, Barbara L., "Developing an IAQ Action Plan," BOM Magazine, BOMA, February, 1996, pg. 70-76.

Wallace, L. et. al. "Personal Exposure to Volatile Organic Compounds." *Env. Res*. Vol. 35 (1984), 193-211.

Weill, H. and J.M. Hughes, "Asbestos as a Public Health Risk: Disease and Policy," *Annual Review of Public Health*, 7: (1986), 171-92.

Wellford, B.W. "Mitigation of Indoor Radon Using Balanced Mechanical Ventilation Systems." *Managing Indoor Air for Health and Energy Conservation*, (1986), 602.

Weschler, Charles J. "Ozone Impact on Public Health' Concentrations from Indoor Exposure to Ozone and Products of Ozone Initiated Chemistry" National Institute of Health. June, 2006.

Wheeler, Arthur E. "Office Building Air Conditioning to Meet Proposed ASHRAE Standard 62-1981 R." *Engineering Solutions to Indoor Air Problems*. (1988), 99-107.

White, Bruce, "Acceptable Fungi & Bacteria Levels. Who Sets the Standards?" ASHRAE IAQ Applications, Fall, 2003

Wilson, Cynthia, "Chemical Intolerance: Hallmark Feature of MCS and Porphyria," Our Toxic Times, Chemical Injury Information Network, October, 1996, pgs. 1-3.

Wolf, James E., "Engineering, Economics, Energy, and the Environment." HPAC Engineering, March, 2001, Pg. 11.

Woods, J.E. and A.K. Goodwin, "Glass Fiber emissions from HVAC Ductwork: A Review of the Literature," Healthy Buildings/IAQ '97 Proceedings, pgs. 587-592.

Woods, James E., "Underfloor and Conventional Air Distribution Systems During Occupancy: A Status Report." *ASHRAE Journal*, May 2, 2003.

Woods, James E., "Comparative Evaluation of Underfloor and conventional Air Distribution Systems: A Literature Review." National Energy Mgmt. Institute, September. 23, 2002.

World Health Organization, "The Right to Healthy Indoor Air." Report on a WHO meeting, May 15-17, 2000.

Wright, Walter G.; Irby, Stephanie M., "The Transactional Challenges Posed by Mold: Risk Management and Allocation Issues." Arkansas Law Review., Vol. 56, #2, 2003

Zegel, William C., "New Technology: May the Force Be With Us," *EM Magazine*, AWMA, October, 1997, pg. 4.

INVESTIGATION FORMS

INTERVIEW FORM

PERSON INTERVIEWED _____ DATE __/__/__

COMPLAINT ON FILE   [ ] YES   [ ] NO

BUILDING/AREA WHERE PERSON WORKS _____

INTERVIEWER _____

----

1. How long have you worked in building? _____ In area? _____
   How much time elapsed before symptoms started? _____

2. Do you have a history of related medical problems (allergies, asthma, respiratory ailments, hayfever, mygrains, eye irritation, eczema, dermatitis, etc.)? If yes, when do they occur?

3. Is any medication taken related to this medical problem?

      [ ] Prescribed   [ ] Over the counter

4. Symptoms experienced and estimated duration:

	0-24 Hrs	< 1 wk	1-4 wks	4 wks
Headache				
Eye irritation				
Nose irritation				
Throat irritation; upper respiratory				
Dry mouth				
Backache				
Shortness of breath				
Chest pains				
Nausea				
Flu-like symptoms				
Fever				
Fatigue, malaise lethargy				
Drowsiness				
Dizziness or faintness				
Difficulty concentrating				
Skin dryness, rash, irritation				
Too hot				
Too cold				

Describe symptoms checked above in more detail:

5. Symptom patterns:

Symptoms occur   [ ] intermittently   [ ] continually

If intermittent,

How often # _____ / (day, week, month) [circle one]

How long do symptoms last (several minutes, several hours, all day, all week)

What months of the year have the symptoms been experienced? (Circle all that apply.)

J   F   M   A   M   J   J   A   S   O   N   D

What days of the week are they most likely to appear? (Circle all that apply.)

S   M   T   W   T   F   S

What time of the day are they most apt to appear? (Circle all that apply.)

All the time   Any time   A.M.   P.M.

Do symptoms vary in intensity?   [ ] Yes [ ] No

If yes, when is problem greatest? _____

Do any of the following apply?

Wears contact lenses   [ ] Yes [ ] No
Operates visual display terminal   [ ] Yes [ ] No
  at least 10% of the day

Operates copiers at least   [ ] Yes [ ] No
  10% of the day
Engages in intensive paper handling,   [ ] Yes [ ] No
  especially carbonless sensitive paper
Uses special office equipment   [ ] Yes [ ] No
  If yes, specify _____

Are symptoms experienced away from work?   [ ] Yes [ ] No

  Previous work locations?   [ ] Yes [ ] No
  At home?   [ ] Yes [ ] No

6. What are the weather conditions when your symptoms are most apt to appear? Or, are worst?

   [ ] Calm, mild   [ ] Windy   [ ] Cold
   [ ] Rainy, stormy   [ ] Hot, humid [ ] Dry

7. Are there any specific work activities you engage in just prior to experiencing these symptoms? Are they more apt to happen in a certain work area?

8. How would you describe conditions around your work area? (Check terms, or terms similar to those, used by the person.)

	stuffy		too smokey		too dry
	too drafty		too much glare		too bright
	too humid		too much noise		poor light
	feet too cold		back too cold		back too hot

9. Have you noticed any unpleasant odor(s)? Describe:

10. Is smoking allowed in the work area?   [ ] Yes [ ] No

    Do you smoke?   [ ] Yes [ ] No

    Are you bothered by smoke?   [ ] Yes [ ] No

11. Have you sought medical attention related to the symptoms?
    If yes, describe: [ ] Yes [ ] No

12. What do you think causes your symptoms?

Signatures: Investigator _____ Date __/__/__

             Interviewee _____ Date __/__/__
                        (optional)

PRELIMINARY ASSESSMENT FORM

BUILDING_____ SPECIFIC AREA(S) _____

DATE OF INVESTIGATION _____

INVESTIGATOR _____

A. NATURE AND SCOPE OF COMPLAINTS

(Summary of interview information)

Number interviewed with complaints _____

Number interviewed that have no complaints _____

Symptoms evidenced (indicate number in box):

Frequency by Time Periods

	0-24 Hrs	< 1 wk	1-4 wks	4 wks
Headache				
Eye irritation				
Nose irritation				
Throat irritation; upper respiratory				
Dry mouth				
Backache				
Shortness of breath				

	0-24 Hrs	< 1 wk	1-4 wks	4 wks
Chest pains				
Nausea				
Flu-like symptoms				
Fever				
Fatigue, malaise lethargy				
Drowsiness				
Dizziness or faintness				
Difficulty concentrating				
Skin dryness, rash, irritation				
Too hot				
Too cold				

Describe areas or activities that seem to be associated with increased episodes:

Discuss any distinctive patterns experienced by more than one person for a particular symptom:

Symptom   No. of People   Pattern
                                   (Time of day, weather,
                                        activity, area)

Is smoking allowed in the area of concern?   [ ] Yes [ ] No

Have odors been detected?   [ ] Yes [ ] No
Describe odor and indicate number reporting it:

Indicate number who described the work area using any of following
terms:

	stuffy		too smokey		too dry
	too drafty		too much glare		too bright
	too humid		too much noise		poor light
	feet too cold		back too cold		back too hot

B. BACKGROUND ASSESSMENT

(A walk through inspection of problem areas and HVAC supply
equipment; no measurements.)

OBSERVATIONS: Time: __:__ A.M.
                                                       __:__ P.M.

Temperature in area (thermostat reading is optional)_____°F

     Does it seem   [ ] too hot [ ] too cold [ ] drafty?

Humidity (no measurement)   [ ] too moist   [ ] too dry

Has there been any recent:

	painting		carpet installation		pesticides used
	cleaners used		new construction		new furniture
	wall covering		draperies		other

Is there any evidence of:

	water damage		excessive noise		poor lighting
	mold growth		dirt near ducts		glare

Elaborate on areas checked _____

_____

_____

Check basement area or floor on grade for proper drainage and existing or potential leaks. Note indicators _____

_____

Note any excessive dust/particles _____

_____

Describe noticeable odors and suggest possible sources: _____

_____

Check for proper storage and use of cleaning agents, pesticides or special supplies, such as photographic supplies, hazardous materials _____

_____

Identify any special equipment that may be a potential pollutant source
(copiers, laser printers) _____

_____

Describe any episodic or unusual events, such as roof leaks

Date	Event	Comments

5. Describe any previous IAQ problems and investigations.

_____

_____

6. Describe any actions taken to date to remedy current problem.

_____

_____

DESIGN, CONSTRUCTION

1. Year built _____ original sq.ft. _____

   Site orientation and adjacent land use _____

_____

   Basic construction _____

Type of windows _____ Do they open? _____

Number of stories _____ Basement? [ ] Yes [ ] No

Addition(s) _____ sq.ft. _____

State original design: occupant level _____ and
functional purposes _____

If different from original construction, describe structural changes

_____

_____

Cite any changes in functions/programs _____

_____

Current occupant level more than planned?   [ ] Yes [ ] No

Does this area or a nearby area have:

	research labs		cigarette smoking		copiers
	motor vehicles		animals		garbage
	graphics mat.		other _____

Describe any other external or attached pollution sources
(garages, loading docks, roads, adjacent building's exhaust)

_____

2. Modifications and major maintenance:

	Year	Comment
New carpet		
Equipment chgs.		
Furniture		
Wall covering		
Plastics		
Painting		
Bldg. function		
Other _____		

3. Envelope type

Walls _____

Floors _____

Roof, ceiling _____

Does the building have sprayed or foamed insulation? _____

If yes, when was it applied? _____

- -

HVAC SYSTEM

NOTE: The following HVAC section should be completed by a person, who is knowledgeable about proper operation and maintenance of HVAC systems.
- -

Is building served by one HVAC system?   [ ] Yes   [ ] No

If portions of the building are served by different units, describe which units serve the areas of concern.

_____

_____

Heating:

Does facility have an on-site boiler?   [ ] Yes   [ ] No
If yes, what is approximate age of boiler? _____

What is heating fuel source? (Check all that apply.)

    [ ] natural gas   [ ] coal   [ ] electricity   [ ] oil #___

Is any auxilliary heat used in area?   [ ] Yes   [ ] No
If yes, describe _____

_____

What is general condition of the boiler? _____

_____

Cooling:

Type of cooling used:
     [ ] none   [ ] central plant   [ ] zone   [ ] individual unit

If central plant, what are the chilled water temperature settings?

_____

Ventilation:

Is there a central ventilation system?   [ ] Yes   [ ] No

Are windows operable?   [ ] Yes   [ ] No

Do other areas of the building share the same conditioned air as the area
where symptoms have occurred?   [ ] Yes   [ ] No

If yes, list those areas that you think warrant further investigation

_____

_____

Indicate type of ventilation system (VAV, dual duct) _____

_____

Where are air intakes? Are they unobstructed? Are they functioning
properly? _____

List any outside sources of contamination where emissions may be entering the building ventilation system. _____

How does exhaust leave the building? _____

Are there processes or activities in the building that may serve as a contaminant source?   [ ] Yes   [ ] No

If yes, are they vented directly?   [ ] Yes   [ ] No

What is current outside air setting? _____

Are outside air dampers on air handling units fixed in the closed position?   [ ] Yes   [ ] No

Or, are units not providing the outside air for another reason?
                                                              [ ] Yes   [ ] No

If yes, what? _____

Are negative pressure conditions creating infiltration of contaminated air?
                                                              [ ] Yes   [ ] No

Do drain pans have proper inclination for drainage?
                                                              [ ] Yes   [ ] No

Are the insides of the ductwork clean?   [ ] Yes   [ ] No

Filtration:

Describe type of filtration used, its location in the system, and any scheduled filter maintenance.

Are filters serviced and accessible?   [ ] Yes   [ ] No
Filter gages installed and operative?   [ ] Yes   [ ] No
Are filters properly mounted, gasketed, and sealed?   [ ] Yes   [ ] No

Humidification/Dehumidification:

Is humidifcation equipment used?   [ ] Yes   [ ] No

If yes, is it central? _____ Or local? _____

Circle type: steam injection   air washer   water spray

Is dehumidification equipment used?   [ ] Yes   [ ] No

  If yes, is it central? _____ Or local? _____

Appendix E

GUIDANCE CHECKLIST FOR DESIGN DOCUMENTATION

Anne Hayner, past editor of *Engineered Systems* noted that "A design that provides high-quality indoor air may not be enough. Regardless of the system type, the designer must now prepare to prove compliance with the standards so clearly, that there will be no doubt in the minds of a jury of butchers, bakers, and candlestick makers." Another way of saying that is Document! Document! Document! Thus, the following listing is provided to help the design team with a helpful checklist of the areas to provide documentation to the building operation team. In turn, the building owner should expect and demand that such documentation be provided to the building operating team. This will assure that the building operating and maintenance staff is properly informed as to the intent and purpose and long term support needs of the building components. This will more fully enable them to operate and maintain the facility in a manner that maximizes performance and air quality acceptability.

- Indoor design targets for key environmental factors for significant building segments, by season and by operating conditions, such as off-hours or summer.

- Outdoor design conditions, including seasonality, prevailing wind conditions, and any local anomalies.

- Occupant densities and space usage patterns by major areas or usage type of space in the facility. Special attention should be paid to off-hour and part-load conditions that would impact the occupant load.

- Load calculations based upon actual existing occupancy loads, lighting levels, and other building-related thermal characteristics.

- Any special thermal, moisture, or contaminant sources which would affect the needs of the occupants of the space from a thermal or ventilation standpoint. Include MSDS and outgassing data on specific constituents of concern.

- Assumptions and design path selected for the development and establishment of ventilation levels by building segment and occupancy densities. This is to include documentation as the procedures selected and the calculations deriving the ventilation rates during the assumed or documented load conditions.

- Air cleaning equipment selected, removal efficiencies, and rationale for selection.

- Minimum outdoor air flows and minimum/maximum supply air flows determined for each significant zone and air handler.

- Documentation outdoor air quality assumptions along with design plans for o/a quality modification.

- Ventilation effectiveness assumptions and actions taken to maximize ventilation delivery.

- Applicable codes, standards, and practices.

- Narrative of operation protocol and intent both during full occupancy as well as part load, including set points and control response strategy.

- Narrative concerning the control tactics specifically devoted to moisture and humidity control.

- HVAC controls data such as sequence descriptions, maintenance and calibration data.

- Air balance information.

- Commissioning reports and related affirmation of operation.

- Maintenance schedules and recommendations.

- Complete drawings and schedules.

- Full Operating and Maintenance Manual with full description of all equipment and full recommendations with appropriate checklist for ongoing preventative maintenance of mechanical equipment.

- Guidance documents, standards, and codes which apply to the above documentation.

INDEX